THIRD EDITION

Nuclear Radiation with Computers and Calculators

Experiments Using a Radiation Monitor with
a Vernier LabPro® or CBL 2™

John E. Gastineau

Vernier Software & Technology
13979 S.W. Millikan Way • Beaverton, OR 97005-2886
Toll Free (888) 837-6437 • (503) 277-2299 • FAX (503) 277-2440
info@vernier.com • www.vernier.com

THIRD EDITION

Nuclear Radiation with Computers and Calculators

Experiments Using a Radiation Monitor with
a Vernier LabPro or CBL 2

Radiation experiments using Vernier Software & Technology products with Macintosh®
and Windows® computers and Texas Instruments graphing calculators for collecting,
displaying, graphing, and analyzing radiation data

The terms CBL, CBL 2, Calculator-Based Laboratory, TI-GRAPH LINK, and TI Connect are either registered trademarks, trademarks, or copyrighted by Texas Instruments, Inc. Vernier LabPro and Logger *Pro* are registered trademarks of Vernier Software & Technology. Macintosh, Mac OS, Quicktime, and Apple II are registered trademarks of Apple Computer, Inc. Windows, Microsoft, MS Word, Windows 2000, Windows ME, Windows NT, Windows XP, Excel, and MS-DOS are registered trademarks of Microsoft Corporation in the United States and/or other countries

Published by
Vernier Software & Technology
13979 S.W. Millikan Way
Beaverton OR 97005-2886
Toll Free 888-837-6437
(503) 277-2299
FAX (503) 277-2440
info@vernier.com
www.vernier.com

ISBN 1-929075-33-2
Third Edition
First Printing
Printed in the United States of America

About the Author

John E. Gastineau earned a Ph.D. in Physics in 1986 from the University of Wisconsin, Madison, doing experimental work in atomic collisions. As a university-level physics instructor for ten years, his primary teaching interest has been the introductory physics course. He used microcomputer-based lab tools and simulations extensively in his lecture-lean teaching, and has given invited talks on their use at both national and international physics and science education meetings.

John is Staff Scientist for Vernier Software & Technology. He is an author of *Physics with Calculators, Physics with Computers, Real-World Math with Computers*, and *Nuclear Radiation for Computers and Calculators*, published by Vernier Software & Technology, as well as *Real-World Math with the CBL2 and LabPro* and *College Physics for the CBL and TI-86*, published by Texas Instruments. An avid whitewater kayaker and mountain biker, he is also a member of the Mt. Hood Ski Patrol, patrolling year-round on a snowboard.

Contents

Experiments for Computer

Experiments for Calculator

Appendices

Preface

Radiation and nuclear decay may be taught as a part of a chemistry, physics or environmental science course. This book is designed to be used as a special unit in such a course. Computers and calculators are an integral and indispensable equipment component in a science laboratory, and these experiments provide a way to offer hands-on radiation experiments using this technology.

This book contains six student experiments using Logger *Pro* (3.3 or newer) for Windows and Macintosh computers[1]. A Vernier LabPro is used as the interface. The same experiments can also be performed using graphing calculators from Texas Instruments. Either a Vernier LabPro or a TI CBL 2 interface can be used with the supplied calculator software called DataRad. The original CBL™ interface is not supported in these experiments.

The experiments may be performed using either a Vernier Radiation Monitor or a Student Radiation Monitor. Both units are Geiger detectors, detecting the presence of ionizing radiation without measuring energy. The slightly more expensive Radiation Monitor is sensitive to alpha particles, while the Student Radiation Monitor is not. Otherwise, the two sensors are equivalent for these experiments.

This book has been revised to include instructions for Logger *Pro* 3 software. If you are using an earlier version of Logger *Pro*, instructions for all experiments can be found in the Word Files for Older Book folder on the book CD.

Following each student experiment, there is an extensive Teacher Information section with sample results, answers to questions, directions for preparing equipment, and other hints regarding the planning and implementation of a particular experiment.

Experiments in this book can be used unchanged or they can be modified using the word-processing files provided on the computer disks. Students will respond differently to the design of the experiments, depending on teaching styles of their teachers, math background, previous experience using computers, and the scope and level of the physics or chemistry course.

In writing these experiments, I included more detailed instructions than some teachers would prefer. I also included more questions than some teachers would want to have their students answer. I realize that some teachers like to give very minimal instructions to their students in the laboratory. My thinking is that the extra information would be helpful to novice students (and teachers), and that other teachers can easily edit the extra text. It is much easier and faster to remove words from the labs than it is to create new ones.

Here are some ways to use the experiments in this book:

- Unchanged. You can photocopy the student sheets, distribute them, and have the experiments done following the procedures as they are printed. Many students will be more comfortable if most of the computer steps used in data collection and analysis are included in each experiment.
- Slightly modified. The CD accompanying the book is for this purpose. Before producing student copies, you can change the directions a little to make them fit your particular teaching circumstances.
- Extensively modified. This, too, can be accomplished using the accompanying CD. Some teachers will want to decrease the degree of detail in student instructions.

---

[1] To upgrade your copy of Logger *Pro* to the current version, go to www.vernier.com. On that page, click on the link to Downloads.

It is **important** for teachers to read the information presented in the appendixes in the back of the book. There is valuable information here that can help make you more comfortable with your initial use of the Logger *Pro* software, the interfaces, and Vernier probes. Here is a short summary of the information available in each appendix:

- Appendix A has detailed information on the DataRad program for TI Graphing Calculators.
- Appendix B tells you how to use the word-processing files on the CD-ROM.
- Appendix C describes equipment from Vernier for performing radiation experiments.

I thank Chris and Dave Vernier, Rick Sorensen, Dan Holmquist, John Wheeler, Ian Honohan, and Garth Upshaw of Vernier Software & Technology for the advice and assistance they provided.

JEG
jgastineau@vernier.com
April 2004

α, β, and γ

Nuclear radiation can be broadly classified into three categories. These three categories are labeled with the first three letters of the Greek alphabet: α (alpha), β (beta) and γ (gamma). Alpha radiation consists of a stream of fast-moving helium nuclei (two protons and two neutrons). As such, an alpha particle is relatively heavy and carries two positive electrical charges. Beta radiation consists of fast-moving electrons or positrons (an antimatter electron). A beta particle is much lighter than an alpha, and carries one unit of charge. Gamma radiation consists of photons, which are massless and carry no charge. X-rays are also photons, but carry less energy than gammas.

After being emitted from a decaying nucleus, the alpha, beta or gamma radiation may pass through matter, or it may be absorbed by the matter. You will arrange for the three classes of radiation to pass through nothing but a thin layer of air, a sheet of paper, and an aluminum sheet. Will the different types of radiation be absorbed differently by the air, paper and aluminum? The question can be answered by considering which radiation type will interact more strongly with matter, and then tested by experiment.

In this experiment you will use small sources of alpha, beta, and gamma radiation. *Follow all local procedures for handling radioactive materials.*

OBJECTIVES

- Develop a model for the relative absorption of alpha, beta, and gamma radiation by matter.
- Use a radiation counter to measure the absorption of alpha, beta, and gamma radiation by air, paper, and aluminum.
- Analyze count rate data to test for consistency with your model.

MATERIALS

computer	Polonium-210 0.1μC alpha source
Vernier computer interface	Strontium-90 0.1μC beta source
Logger *Pro*	Cobalt-60 1μC gamma source
Vernier Radiation Monitor or	paper sheet
Student Radiation Monitor	aluminum sheet, about 2 mm thick

PRELIMINARY QUESTIONS

1. Most nuclear radiation carries energy in the range of a few million electron volts, or MeV ($1 \text{ MeV} = 10^6 \text{ eV} = 1.6 \times 10^{-13}$ J), regardless of its type (alpha, beta, or gamma). This means that more massive particles generally travel more slowly than light particles. Make a preliminary guess as to which radiation type will in general interact most strongly with matter, and therefore would be most strongly absorbed as it passes through matter. Consider electrical charge, mass and speed. Explain your reasons.

2. Which radiation type do you predict would interact, in general, least strongly with matter, and so be less absorbed than others? Why?

3. Which radiation type do you predict would have an intermediate level of interaction with matter? Why?

4. You will be using paper and aluminum sheet metal as absorbers for the radiation. Which material has the greatest areal density (that is, a density per unit area, which could be measured in g/cm^2), and so would present more matter to the passing radiation? Which material would have less?

5. Is your radiation monitor sensitive to all three types of radiation? How can you tell? Devise a test and carry it out. If your radiation monitor does not detect one form of radiation, then you will be able to compare the absorption of the remaining two types.

PROCEDURE

1. Connect the radiation monitor to DIG/SONIC 1 of the computer interface.

2. Prepare the computer for data collection by opening the file "01 Alpha Beta Gamma" from the *Nuclear Radiation w Computers* folder of Logger *Pro*.

3. If you are using the Radiation Monitor (brown plastic case with meter) place the source near the metal screen, and when using an absorber, place the absorber between the source and the screen. If you are using the Student Radiation Monitor (black plastic case with no readout), place the source near the Geiger tube window on the underside, and when using an absorber, place the absorber between the source and the window. In either case, use approximately the same position for the sources each time, with and without an absorber. The sources are usually mounted in small plastic discs, with the most radiation emitted from the underside of the disc.

   Begin with no source, to determine the background count rate. Move all sources away from the monitor. Click ▶ Collect to begin collecting data. While it may appear as if data collection did not start, Logger *Pro* is collecting data. Wait 50 s for the number of counts to appear in the meter. Record the number of counts in the no-source row of your data table, no shielding.

4. Using no absorber, place the beta source near the appropriate region of your radiation monitor, with the underside of the disc facing the monitor. Click ▶ Collect to begin collecting data. Wait for Logger *Pro* to complete data collection. Record the number of counts in the beta row of your data table, no shielding.

5. Place a single sheet of paper between the beta source and the monitor, and measure the counts as before. Take care to keep the source in the same position with respect to the radiation monitor. Record the count rate in the appropriate place.

6. In a similar manner, record the counts for the following used as absorbers for each of the three sources:

   a. a single sheet of paper
   b. a single sheet of aluminum

   Record each count in your data table.

DATA TABLE

Counts in 50 s interval			
	no shielding	shielding	
source		paper	Al sheet
none			
alpha			
beta			
gamma			

ANALYSIS

1. Compare the no-source, or background, count with the no-absorber counts for the sources. Is the background count number a significant fraction of the counts from the sources? Do you need to consider a correction for the background counts?

2. Inspect your data. Does the count rate appear to follow your initial guesses for the relative absorption of the various types of radiation by matter? Be specific, considering which source should be the most penetrating (least interacting), and which absorber is more difficult to penetrate.

3. X-rays are photons, just like gamma rays. X-rays carry lower energy, however, and so historically received a different name. If you have had an X-ray film picture of your teeth taken by a dentist, the dentist probably placed a lead-lined apron on your chest and lap before making the X-ray. What is the function of the lead apron? Support any assertion you make from your experimental data.

EXTENSIONS

1. If you were presented with a safe, but unknown, radiation source, and assuming that it emitted only one type of radiation, devise a test that would allow you to tentatively identify the type of radiation as primarily alpha, beta, or gamma. Write instructions for another student to follow in performing the test.

2. Your monitor detected some radiation even without a source present. Devise a method to correct for this background radiation. Do the corrected data still agree with your prediction?

α, β, and γ

Nuclear radiation can be broadly classified into three categories. These three categories are labeled with the first three letters of the Greek alphabet: α (alpha), β (beta) and γ (gamma). Alpha radiation consists of a stream of fast-moving helium nuclei (two protons and two neutrons). As such, an alpha particle is relatively heavy and carries two positive electrical charges. Beta radiation consists of fast-moving electrons or positrons (an antimatter electron). A beta particle is much lighter than an alpha, and carries one unit of charge. Gamma radiation consists of photons, which are massless and carry no charge. X-rays are also photons, but carry less energy than gammas.

After being emitted from a decaying nucleus, the alpha, beta or gamma radiation may pass through matter, or it may be absorbed by the matter. You will arrange for the three classes of radiation to pass through nothing but a thin layer of air, a sheet of paper, and an aluminum sheet. Will the different types of radiation be absorbed differently by the air, paper and aluminum? The question can be answered by considering which radiation type will interact more strongly with matter, and then tested by experiment.

In this experiment you will use a small sources of alpha, beta, and gamma radiation. *Follow all local procedures for handling radioactive materials.*

OBJECTIVES

- Develop a model for the relative absorption of alpha, beta, and gamma radiation by matter.
- Use a radiation counter to measure the absorption of alpha, beta, and gamma radiation by air, paper, and aluminum.
- Analyze count rate data to test for consistency with your model.

MATERIALS

TI Graphing Calculator
LabPro or CBL 2
DataRad calculator program
Vernier Radiation Monitor or
   Student Radiation Monitor

Polonium-210 0.1μC alpha source
Strontium-90 0.1μC beta source
Cobalt-60 1μC gamma source
paper sheet
aluminum sheet, about 2 mm thick

PRELIMINARY QUESTIONS

1. Most nuclear radiation carries energy in the range of a few million electron volts, or MeV (1 MeV = 10^6 eV = 1.6×10^{-13} J), regardless of its type (alpha, beta, or gamma). This means that more massive particles generally travel more slowly than light particles. Make a preliminary guess as to which radiation type will in general interact most strongly with matter, and therefore would be most strongly absorbed as it passes through matter. Consider electrical charge, mass and speed. Explain your reasons.

2. Which radiation type do you predict would interact, in general, least strongly with matter, and so be less absorbed than others? Why?

3. Which radiation type do you predict would have an intermediate level of interaction with matter? Why?

4. You will be using paper and aluminum sheet metal as absorbers for the radiation. Which material has the greatest areal density (that is, a density per unit area, which could be measured in g/cm^2), and so would present more matter to the passing radiation? Which material would have less?

5. Is your radiation monitor sensitive to all three types of radiation? How can you tell? Devise a test and carry it out. If your radiation monitor does not detect one form of radiation, then you will be able to compare the absorption of the remaining two types.

PROCEDURE

1. Connect the radiation monitor to DIG/SONIC 1 of the LabPro or to DIG/SONIC of the CBL 2. Use the black link cable to connect the TI graphing calculator to the interface. Firmly press in the cable ends.

2. Turn on the calculator and start the DataRad program. Press (CLEAR) to reset the program.

3. Prepare the DataRad program for this experiment.

    a. Select SETUP from the main screen.
    b. Select SET INTERVAL from the SETUP MENU.
    c. Select SET INTERVAL from the INTERVAL SETTINGS menu.
    d. Enter "50" as the count time interval in seconds. Always complete number entries with (ENTER).
    e. Select OK from the INTERVAL SETTINGS menu.
    f. Select SINGLE INTERVAL from the SETUP MENU.

4. If you are using the Radiation Monitor (brown plastic case with meter) place the source near the metal screen, and when using an absorber, place the absorber between the source and the screen. If you are using the Student Radiation Monitor (black plastic case with no readout), you will place the source near the Geiger tube window on the underside, and when using an absorber, place the absorber between the source and the window. In either case, use approximately the same position for the sources each time, with and without an absorber. The sources are usually mounted in small plastic discs, with the most radiation emitted from the underside of the disc. Begin with the no source, to determine the background count rate.

    a. Move all sources away from the monitor.
    b. Select START from the main screen to begin collecting data.
    c. Wait 50 seconds for the calculator to complete the data collection interval. Record the number of counts/minute in the no-source row of the data table, no shielding.
    d. Return to the main screen by pressing (ENTER).

5. Using no absorber, place the beta source near the appropriate region of your radiation monitor, with the underside of the disc facing the monitor. As you did before, select START to begin collecting data. Wait 50 seconds for the calculator to complete the data collection interval. Record the number of counts/minute in the beta row of the data table, no shielding.

6. Place a single sheet of paper between the beta source and the monitor, and measure the counts as before. Take care to keep the source in the same position with respect to the radiation monitor. Record the count rate in the appropriate place.

7. In a similar manner, record the counts for the following used as absorbers for each of the three sources. Record each count rate in your data table.

    a. a single sheet of paper
    b. a single sheet of aluminum

    Record each count rate in your data table.

DATA TABLE

Counts in 50 s interval			
	no shielding	shielding	
source		paper	Al sheet
none			
alpha			
beta			
gamma			

ANALYSIS

1. Compare the no-source, or background, count with the no-absorber counts for the sources. Is the background count number a significant fraction of the counts from the sources? Do you need to consider a correction for the background counts?

2. Inspect your data. Does the count rate appear to follow your predictions for the relative absorption of the various types of radiation by matter? Be specific, considering which source should be the most penetrating (least interacting), and which absorber is more difficult to penetrate.

3. X-rays are photons, just like gamma rays. X-rays carry lower energy, however, and so historically received a different name. If you have had an X-ray film picture of your teeth taken by a dentist, the dentist probably placed a lead-lined apron on your chest and lap before making the X-ray. What is the function of the lead apron? Support any assertion you make from your experimental data.

EXTENSIONS

1. If you were presented with a safe, but unknown, radiation source, and assuming that it emitted only one type of radiation, devise a test that would allow you to tentatively identify the type of radiation as primarily alpha, beta, or gamma. Write instructions for another student to follow in performing the test.

2. Your monitor detected some radiation even without a source present. Devise a method to correct for this background radiation. Do the corrected data still agree with your prediction?

TEACHER INFORMATION

α, β, and γ

1. The absorption model students develop in the first few questions is valid only in broad terms. The actual penetration of radiation can depend strongly on the energy of the particles, which is not considered in the simple reasoning applied here.

2. The Student Radiation Monitor (black case with no meter, model SRM-BTD) is not sensitive to alpha radiation, and so will not respond at all to the alpha source. Students doing this activity with the Student Radiation Monitor will thus not be able to study the relative absorption of alphas. The Radiation Monitor (brown case with analog meter, model RM-BTD) is sensitive to all three types of radiation.

3. The three sources are commonly sold as small plastic discs, with the radioactive material embedded within the plastic. Note that the alpha source has an open window on the underside of the disc. Because alphas are so strongly absorbed, this open window must face the detector during the experiment.

4. Since the alpha source, polonium-210, has a half-life of 138 days, it must be replaced regularly. A source purchased years ago will be dead. You can test your source with the Radiation Monitor (but not the Student Radiation Monitor). Measure the background count in a 50 s interval. Place the open side of the source right next to the screen of the monitor, and measure the counts again. You should get at least three times the background count rate for a useable source.

5. The activity does not ask students to correct for background counts, since the count rate with sources will be so large. You may want to ask students to correct their measurements for the background counts.

6. Sources are available from a number of suppliers:

   - Spectrum Techniques, 106 Union Valley Road, Oak Ridge, TN 37830, (865) 482-9937, Fax: (865) 483-0473, www.spectrumtechniques.com.
   - Flinn Scientific, P.O. Box 219, Batavia, IL 60510, (800) 452-1261, Fax: (630) 879-6962, www.flinnsci.com.
   - Canberra Industries, 800 Research Parkway, Meriden, CT 06450, (800) 243-3955 Fax: (203) 235-1347, www.canberra.com.

DATA TABLE

	no shielding	shielding	
		paper	Al sheet
source			
none	15		
alpha	52	16	14
beta	10691	9904	24
gamma	4809	4503	2766

Counts in 50 s interval

Note that these data were taken with a Radiation Monitor. If a Student Radiation Monitor is used, the alpha count rates will be zero.

ANSWERS TO PRELIMINARY QUESTIONS

1. Compared to betas and gammas, alpha particles are most likely to be absorbed by matter. We might expect that the absorption is large because they have the highest electrical charge, and for a given energy, are moving relatively slowly because of their large mass.

2. Compared to alphas and betas, gammas are least likely to be absorbed by matter. Again, we might expect that the absorption is smaller because they have no charge and move at the speed of light.

3. We expect beta radiation to be absorbed at a rate between that of alphas and gammas, since beta rays have less charge and move faster than alphas, and since gammas have no charge.

4. Compared to paper, the aluminum has the greater areal density. As a result, an aluminum sheet should be more strongly absorbing of radiation than a sheet of paper.

5. Answer will depend on the monitor used. The Radiation Monitor is sensitive to all three types; the Student Radiation Monitor is sensitive only to beta and gamma radiation. This can be determined by holding each source in turn by the window of the detector while running in audio mode.

ANSWERS TO ANALYSIS QUESTIONS

1. The count rate with no source (the background count rate) is much smaller than the count rate with a source. As a result, the correction is insignificant.

2. Yes, the data are consistent with predictions. Judging from the gamma data, the aluminum absorbs more than did the paper, which itself absorbed more than the thin layer of air between the source and the monitor. We expect that the alpha radiation would be most strongly absorbed, then beta, and finally gamma radiation would be the least absorbed. The alpha radiation was stopped by even one sheet of paper. The beta radiation was not stopped by the paper, but was stopped by the aluminum. The gamma radiation was strongly attenuated by the aluminum sheet, but passed easily through air and paper.

3. Since a relatively light aluminum sheet absorbed part of the gamma rays, a heavy lead apron should absorb much of the X-ray radiation. The apron thus shields the patient from X-ray exposure to the torso and pelvis.

ANSWERS TO EXTENSIONS

1. To determine the type of radiation (alpha, beta or gamma), first determine the background count rate, then the source count rate with no absorber. Next, place a sheet of paper between the source and the monitor. If the counts are significantly reduced, the source emits alpha particles. If the count rate is not significantly reduced, place a 2 mm thick sheet of aluminum between the source and the monitor. If the count rate goes almost to the background level, the source is emitting beta particles. If the count rate is only reduced, the source is a gamma source.

2. One way to correct for background counts is to measure the background count rate several times to obtain an average value. Subtract this value from each of the other counts, replacing any negative numbers with zero. Some variation is to be expected with small count rates (see the Counting Statistics activity for more information). The background correction does not change any conclusions.

Distance and Radiation

Scientists and health care workers using intense radiation sources are often told that the best protection is distance; that is, the best way to minimize exposure to radiation is to stay far away from the radiation source. Why is that?

A physically small source of radiation, emitting equally in all directions, is known as a point source. By considering the way radiation leaves the source, you will develop a model for the intensity of radiation as a function of distance from the source. Your model may help explain why users of radiation sources can use distance to reduce their exposure.

In this experiment you will use a small source of gamma radiation. Gamma rays are high-energy photons. If your source behaves as a point source, and if the air absorbs little or none of the gamma radiation, then the radiation intensity should be described well by your model. *Follow all local procedures for handling radioactive materials.*

OBJECTIVES

- Develop a model for the distance-dependence of gamma radiation emitted from a point source.
- Use a counter to measure radiation emitted by a gamma source as a function of distance.
- Analyze count rate data in several ways to test for consistency with the model.

MATERIALS

computer
Vernier computer interface
Logger *Pro*

Vernier Radiation Monitor or
  Student Radiation Monitor
Cobalt-60 1 μC source
meter stick

PRELIMINARY QUESTIONS

1. Place your cobalt-60 source on a table. Turn on the radiation monitor to the audio mode, so that it beeps when radiation is detected. (If your monitor has a range switch, set it to the X1 position.) By holding the monitor near the source, determine the most sensitive place on the detector. That is, roughly where inside the monitor case is the radiation being detected?

2. Starting about a meter from your source, slowly move the monitor closer to the source until they nearly touch. How does the beep rate vary with distance from the source? Would you say that the beep rate is proportional to distance from the source? Or is it an inverse relationship?

3. Sketch a qualitative graph of the beep rate *vs.* distance from the source.

4. Suppose a small radiation source (a point source) is placed at the center of two spheres. The spheres are transparent to the radiation. One sphere has a radius *r*, and the other a radius 2*r*. *N* particles leave the source each second and travel outward toward the spheres. How does the number of particles passing through the inner sphere *per unit area* compare to the number

per unit area passing through the outer sphere? Solve this problem by considering the following:

  a. How many total particles pass through the first sphere? How many pass through the second sphere?
  b. How do the surface areas of the two spheres compare?

5. From your answer to the previous question, write down an expression for the intensity of radiation (number of particles passing through a unit area each second) as a function of distance from a point source. Assume that N particles leave the source each second. This expression is your model for the way radiation intensity varies with distance. Record your model in the data table.

6. Is your model consistent with the qualitative sketch you drew in question 3 above?

PROCEDURE

1. Connect the radiation monitor to DIG/SONIC 1 of the computer interface

2. When your source is far from the radiation monitor, the monitor still detects background counts from cosmic rays and other sources. You will need to correct for this background by determining the average count rate with no source near the monitor. Prepare the computer for data collection by opening the file "02a Distance" from the *Nuclear Radiation w Computers* folder. One graph is displayed: counts *vs.* time. The vertical axis is scaled from 0 to 15 counts/interval. The horizontal axis is distance scaled from 0 to 300 seconds. Logger *Pro* will count for ten 30 second intervals. Move all sources at least 2 meters from the radiation monitor, and click ▶ Collect. Wait five minutes for data collection to complete.

3. After Logger *Pro* has finished data collection, click once on the graph to make it active. Notice that the number of counts in each interval varies. This is to be expected since radioactivity is a random process. Click the statistics button on the toolbar to determine the average number of counts in an interval. Record the mean value in the data table.

4. Prepare the computer for data collection by opening the file "02b Distance" from the *Nuclear Radiation w Computers* folder of Logger *Pro*. One graph is displayed: Corrected Radiation (counts/int) *vs.* Distance (m). The vertical axis is scaled from 0 to 300 counts/interval. The horizontal axis is distance scaled from 0 to 0.50 m.

5. Enter your correction for the count rate by modifying a column in the Logger *Pro* data table. To do this, choose Column Options ▶ Corrected Radiation from the Data menu. The Equation field will read "Radiation" – 0. Change the zero to your average background count rate. For example, if your average rate was 7.3, your equation should read "Radiation" – 7.3. Click ☐ Done ☐ to complete the modification.

6. If you are using the Radiation Monitor (brown plastic case) measure all distances from the center of the metal screen, perpendicular to the screen. If you are using the Student Radiation Monitor (black plastic case), stand the case on edge, with the Geiger Tube window near the table. Measure distances from the middle of this window, perpendicular to the window. Place the center of the source 6 cm from the monitor.

7. Click ▶ Collect to begin collecting data. Logger *Pro* will begin counting the number of gamma photons that strike the detector during each 30 second count interval.

8. After at least 30 seconds have elapsed, click the [Keep] button. In the entry field that appears, enter **0.06**, which is the distance in meters from the detector to the center of the source. Complete your entry by clicking [OK]. Data collection will now pause for you to adjust the apparatus.

9. Move the source 0.02 m farther from the source. Click [Continue] to collect more data, and wait 30 seconds.

10. Click [Keep], and enter the new distance of **0.08** meters.

11. In the same way as before, move the source away an additional 0.02 m, click [Continue], wait thirty seconds, and click [Keep]. Enter the distance in meters. Repeat this process until the distance is at least 0.24 m or the counts in one 30 second interval drops below ten.

12. Click [Stop Collection] instead of [Continue] to end data collection.

DATA TABLE

Model expression	
Average background counts	

ANALYSIS

1. Inspect your graph. Does the count rate appear to follow your model?

2. Fit an appropriate function to your data. To do this, click once on your graph to select it, then click the curve fit button [ ]. Select an equation that has the same mathematical form as your model from the equation list, and then click [Try Fit]. A best-fit curve will be displayed on the graph. If your data follow the model, the curve should closely match the data. If the curve does not match well, try a different fit and click [Try Fit] again. When you are satisfied with the fit, click [OK].

3. Print or sketch your graph.

4. From the evidence presented in your two graphs, does the gamma radiation emitted by your source follow your model? Does the relationship seem to fail at larger or smaller distances?

EXTENSIONS

1. Replot your data using a suitable transformation of the *x*-coordinate so that the resulting plot should be linear if the data follow your model. For example, if your model were an inverse-cube function, replot the data using the inverse-cube of the distance values for the horizontal axis. Do your data follow the model well using this test?

2. Why were you instructed to place the source no closer than 0.06 m from the detector? Repeat the experiment, using distances of 0, 0.02, 0.04… out to 0.24 m. Hint: Is the detector a spherical surface?

3. Use a longer counting interval so that you collect at least 300 counts at 0.06 m. Is the agreement with the inverse-square relationship any different? Try a much shorter count interval. How is the resulting graph different? Why?

4. Sometimes the table surface can scatter gamma rays, interfering with data collection. Use a ring stand or other support to hold the sample above the monitor, so that there are no surfaces near the source. Repeat data collection. Do your data agree any better with your model?

Distance and Radiation

Scientists and health care workers using intense radiation sources are often told that the best protection is distance; that is, the best way to minimize exposure to radiation is to stay far away from the radiation source. Why is that?

A physically small source of radiation, emitting equally in all directions, is known as a point source. By considering the way radiation leaves the source, you will develop a model for the intensity of radiation as a function of distance from the source. Your model may help explain why users of radiation sources can use distance to reduce their exposure.

In this experiment you will use a small source of gamma radiation. Gamma rays are high-energy photons. If your source behaves as a point source, and if the air absorbs little or none of the gamma radiation, then the radiation intensity should be described well by your model. *Follow all local procedures for handling radioactive materials.*

OBJECTIVES

- Develop a model for the distance-dependence of gamma radiation emitted from a point source.
- Use a counter to measure radiation emitted by a gamma source as a function of distance.
- Analyze count rate data in several ways to test for consistency with the model.

MATERIALS

TI Graphing Calculator	Vernier Radiation Monitor or
LabPro or CBL 2	Student Radiation Monitor
DataRad calculator program	Cobalt-60 1 μC source
	meter stick

PRELIMINARY QUESTIONS

1. Place your cobalt-60 source on a table. Turn on the radiation monitor to the audio mode, so that it beeps when radiation is detected. (If your monitor has a range switch, set it to the X1 position.) By holding the monitor near the source, determine the most sensitive place on the detector. That is, roughly where in the monitor's case is the radiation being detected?

2. Starting about a meter from your source, slowly move the monitor closer to the source until they nearly touch. How does the beep rate vary with distance from the source? Would you say that the beep rate is proportional to distance from the source? Or is it an inverse relationship?

3. Sketch a qualitative graph of the beep rate *vs.* distance from the source.

4. Suppose a small radiation source (a point source) is placed at the center of two spheres. The spheres are transparent to the radiation. One sphere has a radius r, and the other a radius $2r$. N particles leave the source each second and travel outward toward the spheres. How does the number of particles passing through the inner sphere *per unit area* compare to the number

per unit area passing through the outer sphere? Solve this problem by considering the following:

   a. How many total particles pass through the first sphere? How many pass through the second sphere?
   b. How do the surface areas of the two spheres compare?

5. From your answer to the previous question, write down an expression for the intensity of radiation (number of particles passing through a unit area each second) as a function of distance from a point source. Assume that N particles leave the source each second. This expression is your model for the way radiation intensity varies with distance. Record your model in the data table.

6. Is your model consistent with the qualitative sketch you drew in question 3 above?

PROCEDURE

1. Connect the radiation monitor to DIG/SONIC 1 of the LabPro or to DIG/SONIC of the CBL 2. Use the black link cable to connect the TI graphing calculator to the interface. Firmly press in the cable ends.

2. Turn on the calculator and start the DataRad program. Press ⟨CLEAR⟩ to reset the program.

3. When your source is far from the radiation monitor, the monitor still detects background counts from cosmic rays and other sources. You will need to correct for this background by determining the average count rate with no source near the monitor. Prepare the DataRad program for this experiment by measuring the rate of background events.

   a. Move all sources at least 2 meters from the radiation monitor.
   b. Select SETUP from the main screen.
   c. Select SET INTERVAL from the SETUP MENU.
   d. Select SET INTERVAL from the INTERVAL SETTINGS screen.
   e. Enter "60" as the count time interval in seconds. Always complete number entries with ⟨ENTER⟩.
   f. Select OK from the INTERVAL SETTINGS screen.
   g. Select BACKGROUND CORRECTION from the SETUP MENU.
   h. Select PERFORM NOW from the BACKGROUND CORRECTION screen.
   i. Enter "5" as the number of intervals. The calculator is then ready to count background events for 5*60=300 seconds.
   j. Press ⟨ENTER⟩ to begin counting background events. The remaining time will count down after each interval on the calculator screen.
   k. After 300 seconds, the calculator will display the average background count rate. For the following measurements the count rates will be reduced by the indicated value. Press ⟨ENTER⟩ to continue.

4. Now that the calculator will correct for background events, you are ready to study the relationship between distance and count rate. Prepare the calculator for this new mode by selecting EVENTS WITH ENTRY from the SETUP MENU.

5. If you are using the Radiation Monitor (brown plastic case) measure all distances from the center of the metal screen, perpendicular to the screen. If you are using the Student Radiation Monitor (black plastic case), stand the case on edge, with the Geiger Tube window near the

table. Measure distances from the middle of this window, perpendicular to the window. Place the center of the source 6 cm from the monitor.

6. Select START from the main screen. You will see a new screen from which you will control the counting for each distance between the source and the radiation monitor. Press (ENTER), and the interface will begin counting the number of gamma photons that strike the detector during a 60-second count interval. The screen will display COUNTING...

7. After 60 seconds have elapsed, you will see a prompt. In the entry field that appears, enter "0.06" which is the distance in meters from the detector to the center of the source. Complete your entry by pressing (ENTER).

8. Move the source 0.02 m farther from the source. Press (ENTER) to count for the new distance. When you see the prompt, enter the new distance of 0.08 meters.

9. In the same way as before, move the source an additional 0.02 m out, press (ENTER), and wait for counting to complete. Enter the new distance in meters. Repeat this process until the distance is at least 0.24 m or the rate drops below ten counts per minute.

10. Press (STO▶) to end data collection. You will see a graph of count rate *vs.* distance.

DATA TABLE

Model expression	
Average background counts	
Fit parameters and equation	

ANALYSIS

1. Inspect your graph. Does the count rate appear to follow your model?

2. Fit an appropriate function to your data.

   a. To do this, press (ENTER) to return to the main screen.
   b. Select ANALYZE from the main screen.
   c. Select one of the curve fit relationships from the list on the CURVE FITS screen. Choose the one that matches the mathematical form of your model.
   d. Record the fit equation and parameters in your data table.
   e. Press (ENTER) to see your data with the fitted curve.

3. Print or sketch your graph.

4. From the evidence presented in your two graphs, does the gamma radiation emitted by your source follow your model? Does the relationship seem to fail at larger or smaller distances?

EXTENSIONS

1. Replot your data using a suitable transformation of the *x*-coordinate so that the resulting plot should be linear if the data follow your model. For example, if your model were an inverse-cube function, replot the data using the inverse-cube of the distance values for the horizontal axis. Do your data follow the model well using this test?

2. Why were you instructed to place the source no closer than 0.06 m from the detector? Repeat the experiment, using distances of 0, 0.02, 0.04… out to 0.24 m. Hint: Is the detector a spherical surface?

3. Use a longer counting interval (120 to 300 s, depending on how patient you are!). Is the agreement with the inverse-square relationship any different? Try a much shorter count interval. How is the resulting graph different? Why?

4. Sometimes the table surface can scatter gamma rays, interfering with data collection. Use a ring stand or other support to hold the sample above the monitor, so that there are no surfaces near the source. Repeat data collection. Do your data agree any better with your model?

5. Instead of using one of the built-in curve fits for your data, use the ADD MODEL function on the ANALYZE OPTIONS screen in DataRad to superimpose an equation with exactly the same form as your model on your data. Does this equation fit your data? Is this a better test of your model? Why? To use ADD MODEL you must first enter an equation in the calculator's Y1= field, making use of one or more parameters A, B, C, D or E.

TEACHER INFORMATION

Distance and Radiation

1. Sources are available from a number of suppliers:

   - Spectrum Techniques, 106 Union Valley Road, Oak Ridge, TN 37830, (865) 482-9937, Fax: (865) 483-0473, www.spectrumtechniques.com.
   - Flinn Scientific, P.O. Box 219, Batavia, IL 60510, (800) 452-1261, Fax: (630) 879-6962, www.flinnsci.com.
   - Canberra Industries, 800 Research Parkway, Meriden, CT 06450, (800) 243-3955 Fax: (203) 235-1347, www.canberra.com.

2. Because the radiation monitor detects individual gamma ray arrivals, Poisson statistics apply. The more counts that arrive in a counting interval, the better the precision. The standard error of a count of n is $n^{1/2}$, so do not be surprised to see considerable run-to-run variation in the long distance counts where n is only 10 or 20. To achieve better precision requires larger count numbers, and hence longer count intervals. The computer experiment file provided uses a 30-second counting interval as a compromise between good results and a rapid experiment. Better results will require longer counting intervals.

SAMPLE RESULTS

The computer-based sample data were collected using a Radiation Monitor, while the calculator data were collected using a Student Radiation Monitor.

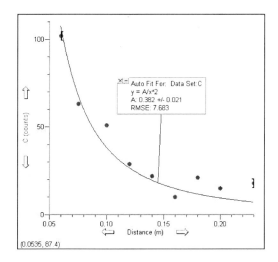

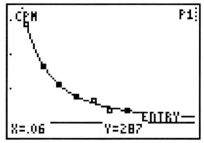
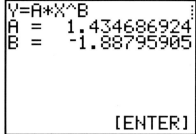

DATA TABLE

Model expression	$I(r) = N/(4\pi r^2)$
Average background counts	8.45

ANSWERS TO PRELIMINARY QUESTIONS

1. The most sensitive place on the Student Radiation Monitor is the clear window on the underside of the monitor. The most sensitive place on the Radiation Monitor is the screen at the end of the monitor case.

2. Since the beep rate increases sharply with decreasing distance, the intensity appears to be an inverse function of distance. It is hard to say from the beeps alone if it is inverse, inverse-square, or otherwise.

3. Sketch should be a decreasing function with distance.

4. The number of particles passing through a unit area decreases as the inverse of the square of the distance from the source. The same number of particles pass through each sphere, but the area of the larger sphere (radius $2r$) is four times the area of the smaller sphere (radius r).

5. $I(r) = N/(4\pi r^2)$, where N is the number of particles leaving the source each second, and $I(r)$ is the number of particles per second per unit area at a distance r from the source. That is, the intensity I is an inverse-square function of distance.

6. Yes, this model is also a decreasing function.

ANSWERS TO ANALYSIS QUESTIONS

1. The count rate falls off rapidly with increasing distance. This is consistent with the inverse-square relationship predicted by the model.

4. The inverse-square function fits the data well, so it appears that the gamma rays do follow the inverse-square law over the range of distances investigated.

ANSWERS TO EXTENSIONS

1. A graph using the inverse-square of the distance from the source for the horizontal axis appears proportional, supporting the inverse-square model.

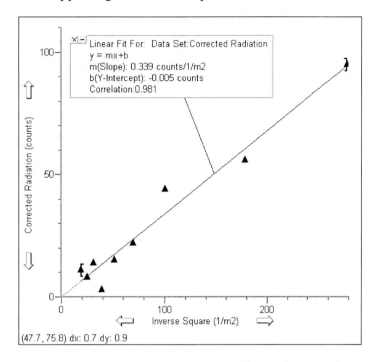

2. At very small distances the entire detector is not a uniform distance from the source, so the effective distance is larger than the source-to-detector center distance. The count rates will then be systematically small. This effect is more significant with the Student Radiation Monitor than with the Radiation Monitor, since the Geiger tube of the former is larger.

3. Longer counting intervals improve the statistics so there is less interval-to-interval variation. As a result the data closely follow an inverse-square function. Shorter count intervals result in more variation, so there is more scatter of points about the inverse-square function.

4. Results will vary. Comparison of different runs is potentially difficult due to counting statistics unless *n* is very large. See Experiment 4 for more information on statistics.

5. (calculator only) Using an inverse-square function of $y1=A/x^2$ and adjusting A so that the curve passes through the first data point produces this graph:

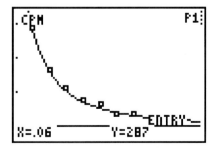

    From the excellent agreement we can conclude that the data support the inverse square model.

Lifetime Measurement

The *activity* (in decays per second) of some radioactive samples varies in time in a particularly simple way. If the activity (R) in decays per second of a sample is proportional to the amount of radioactive material ($R \propto N$, where N is the number of radioactive nuclei), then the activity must decrease in time exponentially:

$$R(t) = R_0 e^{-\lambda t}$$

In this equation λ is the *decay constant*, commonly measured in s^{-1} or min^{-1}. R_0 is the activity at $t = 0$. The SI unit of activity is the becquerel (Bq), defined as one decay per second.

You will use a source called an isogenerator to produce a sample of radioactive barium. The isogenerator contains cesium-137, which decays to barium-137. The newly made barium nucleus is initially in a long-lived excited state, which eventually decays by emitting a gamma photon. The barium nucleus is then stable, and does not emit further radiation. Using a chemical separation process, the isogenerator allows you to remove a sample of barium from the cesium-barium mixture. Some of the barium you remove will still be in the excited state and will subsequently decay. It is the activity and lifetime of the excited barium you will measure.

While the decay constant λ is a measure of how rapidly a sample of radioactive nuclei will decay, the *half-life* of a radioactive species is also used to indicate the rate at which a sample will decay. A half-life is the time it takes for half of a sample to decay. That is equivalent to the time it takes for the activity to drop by one-half. Note that the half-life (often written as $t_{1/2}$) is not the same as the decay constant λ, but they can be determined from one another.

Follow all local procedures for handling radioactive materials. Follow any special use instructions included with your isogenerator.

OBJECTIVES

- Use a radiation counter to measure the decay constant and half-life of barium-137.
- Determine if the observed time-variation of radiation from a sample of barium-137 is consistent with simple radioactive decay.

MATERIALS

computer	Vernier Radiation Monitor or
Vernier computer interface	Student Radiation Monitor
Logger *Pro*	Cesium/Barium-137 Isogenerator
	cut-off paper cup for Barium solution

PRELIMINARY QUESTIONS

1. Consider a candy jar, initially filled with 1000 candies. You walk past it once each hour. Since you don't want anyone to notice that you're taking candy, each time you take 10% of the candies remaining in the jar. Sketch a graph of the number of candies for a few hours.

2. How would the graph change if instead of removing 10% of the candies, you removed 20%? Sketch your new graph.

PROCEDURE

1. Prepare a shallow cup to receive the barium solution. The cup sides should be no more than 1 cm high.

2. Connect the radiation monitor to DIG/SONIC 1 of the computer interface. Turn on the monitor.

3. Prepare the computer for data collection by opening the file "03 Lifetime" from the *Nuclear Radiation w Computers* folder of Logger *Pro*. One graph is displayed: count rate *vs.* time. The vertical axis is scaled from 0 to 1200 counts/interval. The horizontal axis is time scaled from 0 to 30 minutes.

4. Prepare your isogenerator for use as directed by the manufacturer. Extract the barium solution into the prepared cup. Work quickly between the time of solution extraction and the start of data collection in step 6, for the barium begins to decay immediately.

5. Place the radiation monitor on top of or adjacent to the cup so that the rate of flashing of the red LED is maximized. Take care not to spill the solution.

6. Click ▶ Collect to begin collecting data. Logger *Pro* will begin counting the number of gamma photons that strike the detector during each 30 second count interval. Data collection will continue for 30 minutes. Do not move the detector or the barium cup during data collection.

7. After data collection is complete, the ▶ Collect button will reappear. Set the radiation monitor aside, and dispose of the barium solution and cup as directed by your instructor.

DATA TABLE

Average background counts	
fit parameters for Y = A exp (– C*X) + B	
A	
B	
C	
λ (min^{-1})	
$t_{1/2}$ (min)	

ANALYSIS

1. Inspect your graph. Does the count rate decrease in time? Is the decrease consistent with an activity proportional to the amount of radioactive material remaining?

2. Compare your graph to the graphs you sketched in the Preliminary Questions. How are they different? How are they similar? Why are they similar?

3. The solution you obtained from the isogenerator may contain a small amount of long-lived cesium in addition to the barium. To account for the counts due to any cesium, as well as for counts due to cosmic rays and other background radiation, you can measure the background count rate from your data. By taking data for 30 minutes, the count rate should have gone down to a nearly constant value, aside from normal statistical fluctuations. The counts during each interval in the last five minutes should be nearly the same as for the 20 to 25 minute interval. If so, you can use the average rate at the end of data collection to correct for the counts not due to barium. To do this,

   a. Select the data on the graph between 25 and 30 minutes by dragging across the region with your mouse.

   b. Click on the statistics button on the toolbar.

   c. Read the average counts during the intervals from the floating box, and record the value in your data table.

4. Fit an exponential function to the first fifteen minutes of your data. Select the first fifteen minutes with the mouse. Then, click the curve fit button ⌀. Select Natural Exponential from the equation list. Notice that the Natural Exponential fit [y=A*exp(–Ct)+B] includes an additive term B. This term will account for the constant background counts due to non-barium sources. In the number-entry box labeled B value, enter the average background count rate you determined in the previous step, then click ⌐ Try Fit ⌐. A best-fit curve will be displayed on the graph. If your data follow the exponential relationship, the curve should closely match the data. When you are satisfied with the fit, click ⌐ OK ⌐

5. Print or sketch your graph.

6. Record the fit parameters A, B, and C in your data table.

7. From the definition of half-life, determine the relationship between half-life ($t_{1/2}$, measured in minutes) and decay constant (λ, measured in min^{-1}). Hint: After a time of one half-life has elapsed, the activity of a sample is one-half of the original activity.

8. From the fit parameters, determine the decay constant λ and then the half-life $t_{1/2}$.

9. Is your value of $t_{1/2}$ consistent with the accepted value of approximately 2.552 minutes for the half-life of barium-137?

10. What fraction of the initial activity of your barium sample would remain after 25 minutes? Was it a good assumption that the counts in the last five minutes would be due entirely to non-barium sources?

EXTENSIONS

1. How would a graph of the log of the count rate *vs.* time appear? Using Logger *Pro*, Graphical Analysis, or a spreadsheet, make such a graph. Interpret the slope of the line if the data follow a line. Will correcting for the background count rate affect your graph?

2. Repeat your experiment several times to estimate an uncertainty to your decay constant measurement.

3. How long would you have to wait until the activity of your barium sample is the same as the average background radiation? You will need to measure the background count rate carefully to answer this question.

Lifetime Measurement

The *activity* (in decays per second) of some radioactive samples varies in time in a particularly simple way. If the activity (R) in decays per second of a sample is proportional to the amount of radioactive material ($R \propto N$, where N is the number of radioactive nuclei), then the activity must decrease in time exponentially:

$$R(t) = R_0 e^{-\lambda t}$$

In this equation λ is the *decay constant*, commonly measured in s^{-1} or min^{-1}. R_0 is the activity at $t = 0$. The SI unit of activity is the becquerel (Bq), defined as one decay per second.

You will use a source called an isogenerator to produce a sample of radioactive barium. The isogenerator contains cesium-137, which decays to barium-137. The newly made barium nucleus is initially in a long-lived excited state, which eventually decays by emitting a gamma photon. The barium nucleus is then stable, and does not emit further radiation. Using a chemical separation process, the isogenerator allows you to remove a sample of barium from the cesium-barium mixture. Some of the barium you remove will still be in the excited state and will subsequently decay. It is the activity and lifetime of the excited barium you will measure.

While the decay constant λ is a measure of how rapidly a sample of radioactive nuclei will decay, the *half-life* of a radioactive species is also used to indicate the rate at which a sample will decay. A half-life is the time it takes for half of a sample to decay. That is equivalent to the time it takes for the activity to drop by one-half. Note that the half-life (often written as $t_{1/2}$) is not the same as the decay constant λ, but they can be determined from one another.

Follow all local procedures for handling radioactive materials. Follow any special use instructions included with your isogenerator.

OBJECTIVES

- Use a radiation counter to measure the decay constant and half-life of barium-137.
- Determine if the observed time-variation of radiation from a sample of barium-137 is consistent with simple radioactive decay.

MATERIALS

TI Graphing Calculator	Vernier Radiation Monitor or
LabPro or CBL 2	Student Radiation Monitor
DataRad calculator program	Cesium/Barium-137 Isogenerator
	cut-off paper cup for Barium solution

PRELIMINARY QUESTIONS

1. Consider a candy jar, initially filled with 1000 candies. You walk past it once each hour. Since you don't want anyone to notice that you're taking candy, each time you take 10% of the candies remaining in the jar. Sketch a graph of the number of candies for a few hours.

2. How would the graph change if instead of removing 10% of the candies, you removed 20%? Sketch your new graph.

PROCEDURE

1. Prepare a shallow cup to receive the barium solution. The cup sides should be no more than 1 cm high.

2. Connect the radiation monitor to DIG/SONIC 1 of the LabPro or to DIG/SONIC of the CBL 2. Use the black link cable to connect the TI graphing calculator to the interface. Firmly press in the cable ends. Turn on the monitor.

3. Turn on the calculator and start the DataRad program. Press ⌈CLEAR⌉ to reset the program.

4. Prepare the DataRad program for this experiment.

   a. Select SETUP from the main screen.
   b. Select TIME GRAPH from the SETUP MENU.
   c. Select CHANGE TIME SETTINGS from the TIME GRAPH SETTINGS screen.
   d. Enter "30" as the count time interval in seconds. Always complete number entries with ⌈ENTER⌉.
   e. Enter "60" as the number of samples. This setting will give you a 60*30=1800 second (30 minute) data collection time.
   f. Select OK from the TIME GRAPH SETTINGS screen.

5. Prepare your isogenerator for use as directed by the manufacturer. Extract the barium solution into the prepared cup. Work quickly between the time of solution extraction and the start of data collection in step 6, for the barium begins to decay immediately.

6. Place the radiation monitor on top of or adjacent to the cup so that the rate of flashing of the red LED is maximized. Take care not to spill the solution.

7. Select START from the main screen to begin collecting data. The calculator will begin counting the number of gamma photons that strike the detector during each 30-second count interval. Data collection will continue for 30 minutes. Do not move the detector or the barium cup during data collection.

8. As data collection continues a graph will be updated. When collection is complete, a final graph of count rate *vs.* time will appear. Set the radiation monitor aside, and dispose of the barium solution and cup as directed by your instructor.

DATA TABLE

fit parameters for Y = A exp (– B*X) + C	
A	
B	
C	
λ (min^{-1})	
$t_{1/2}$ (min)	

ANALYSIS

1. Inspect your graph. Does the count rate decrease in time? Is the decrease consistent with an activity proportional to the amount of radioactive material remaining?

2. Compare your graph to the graphs you sketched in the Preliminary Questions. How are they different? How are they similar? Why are they similar?

3. The solution you obtained from the isogenerator may contain a small amount of long-lived cesium in addition to the barium. To account for the counts due to any cesium, as well as for counts due to cosmic rays and other background radiation, you can determine the background count rate from your data. By taking data for 30 minutes, the count rate should have gone down to a nearly constant value, aside from normal statistical fluctuations. The counts during each interval in the last five minutes should be nearly the same as for the 20 to 25 minute interval. If so, you can use the average rate at the end of data collection to correct for the counts not due to barium. You will have to do the adjustment outside of the DataRad program.

   a. Use the ⬛▶ cursor key to scan across your graph. The coordinates of the highlighted point are shown on the screen. Determine the lowest count rate in the tail or right-hand region of the graph. Record this value in your data table as the additive constant C.

   b. Press ENTER to return to the main screen.

   c. Select QUIT from the main screen.

   d. Your count rate data are stored in list L2. To subtract the background counts from all the measured count rates, you need to enter L2 – C + 1 STO▶ L2, where C is the constant you determined in step a. This will replace the existing list L2 with the corrected values. The +1 term is necessary to avoid having any of the elements in the L2 list equal to zero, which would cause the exponential fit to fail. To enter L2 on the TI-83, TI-83 Plus, or TI-84 Plus (or any of the first six lists) press 2nd and then the corresponding digit. On the TI-73, access lists by pressing 2nd [STAT]. You will then see a list of available lists. You will need to scroll down using the cursor keys to see all the lists. On other calculators directly enter the list name using the alphanumeric keys.

   e. After you perform the list calculation, your screen will show something like L2 – 8 + 1 →L2, followed by the first part of the actual list contents.

   f. Restart the DataRad program.

4. Now that you have removed the counts from the extra cesium, you can fit an exponential function to the first fifteen minutes of your data. You will use only the first half of the data because the final fifteen minutes now contain largely noise after the background subtraction.

   a. Select ANALYZE from the main screen.

   b. Select SELECT REGION from the ANALYZE OPTIONS screen.

   c. Leave the left bound cursor at the extreme left side of the graph. Press [ENTER] to mark this bound of the selection.

   d. Use the [▶] key to move the cursor to the 15-minute point (900 seconds).

   e. Press [ENTER] to mark this position as the right bound of your selection. The calculator will display a graph showing only the first fifteen minutes of data. Press [ENTER] again to return to the ANALYZE OPTIONS screen.

   f. Select EXPONENT CURVE FIT from the CURVE FITS screen.

   g. Record the fit parameters A and B in your data table.

   h. Press [ENTER] to see your graph with the fitted line.

5. Print or sketch your graph.

6. Press [ENTER], select RETURN TO MAIN SCREEN, and then choose QUIT to leave the DataRad program.

7. From the definition of half-life, determine the relationship between half-life ($t_{1/2}$, measured in minutes) and decay constant (λ, measured in min^{-1}). Hint: After a time of one half-life has elapsed, the activity of a sample is one-half of the original activity.

8. From the fit parameters, determine the decay constant λ and then the half-life $t_{1/2}$.

9. Is your value of $t_{1/2}$ consistent with the accepted value of approximately 2.552 minutes for the half-life of barium-137?

10. What fraction of the initial activity of your barium sample would remain after 25 minutes? Was it a good assumption that the counts in the right side of the graph would be due entirely to non-barium sources?

EXTENSIONS

1. How would a graph of the log of the count rate *vs.* time appear? Using your calculator, Logger *Pro*, Graphical Analysis, or a spreadsheet, make such a graph. Interpret the slope of the line if the data follow a line. Will correcting for the background count rate affect your graph?

2. Repeat your experiment several times to estimate an uncertainty to your decay constant measurement.

3. How long would you have to wait until the activity of your barium sample is the same as the average background radiation? You will need to measure the background count rate carefully to answer this question.

Lifetime Measurement

1. Alert readers may notice that the Preliminary Questions are the same as those in Experiment 27 (Capacitors) of *Physics with Computers*. This duplication is intentional, as both the decay in capacitor potential in an *RC* circuit and radioactive decay are described by exponential functions. You may wish to call your students' attention to this.

2. Sources are available from a number of suppliers:

   - Spectrum Techniques, 106 Union Valley Road, Oak Ridge, TN 37830, (865) 482-9937, Fax: (865) 483-0473, www.spectrumtechniques.com.
   - Flinn Scientific, P.O. Box 219, Batavia, IL 60510, (800) 452-1261, Fax: (630) 879-6962, www.flinnsci.com.
   - Canberra Industries, 800 Research Parkway, Meriden, CT 06450, (800) 243-3955 Fax: (203) 235-1347, www.canberra.com.

3. Detailed directions for preparing the isogenerator are not given because the method varies with manufacturer. You may want to insert the instructions appropriate to your isogenerator at step two of the Procedure.

4. Students often confuse the decay constant parameter λ with the half-life $t_{1/2}$. The decay constant λ is larger for more rapidly decaying elements and has dimensions of time^{-1}, while the half-life has dimensions of time, and is smaller for more rapidly decaying elements. The decay constant λ is equal to the fit parameter C in the Natural Exponential fit of Logger *Pro*. The two parameters can be related in the following manner. After one half-life has elapsed, half of the radioactive nuclei have decayed, and so the activity is also cut in half. From the rate equation we can relate the decay constant to the half life.

$$R = R_0 e^{-\lambda t} \;; \text{at } t = t_{1/2} \text{ we know that } R = \tfrac{1}{2}R_0$$

$$\frac{1}{2}R_0 = R_0 e^{-\lambda t_{1/2}}$$

$$\frac{1}{2} = e^{-\lambda t_{1/2}} \text{. Taking the log of both sides,}$$

$$-\ln 2 = -\lambda t_{1/2}$$

$$t_{1/2} = \frac{\ln 2}{\lambda}$$

There is sufficient information in the student guide to perform this conversion, although some students with weak algebra skills may have difficulty with it. You may choose to work through this step with your students.

5. The cesium-137 in the isogenerator decays to a metastable state of barium. The metastable barium decays with a half-life of 2.552 minutes by gamma emission, making this system an ideal one for studying in the classroom. A 30 minute experimental run covers almost twelve half-lives, so that the observed activity drops to about 0.3% of the initial value.

6. The lifetime obtained depends strongly on the correct subtraction of background (in this case, non-barium) counts. As written, the activity instructions call for a 30-minute data collection period. If time permits, use a 45 or 60 minute period, and measure the count rate for the final 10 or 15 minutes. A longer experiment will ensure that essentially all the barium will have decayed. The sample data shown here yield a lifetime of 2.50 minutes, but if the background value obtained during the last 10 minutes of a 60 minute run is used, the lifetime changes to 2.57 minutes.

7. Many isogenerators allow some cesium to leak through into the barium extract solution. The cesium results in a nearly constant background activity. This background count is often much larger than the environmental background, and the analysis must take it into account. That is why the experiment is written to run for 30 minutes. The final 5 minutes of data can be used to determine the count rate from the combination of cosmic rays and leaked cesium. If you have an isogenerator that does not leak significant amounts of cesium, you may want to shorten the experiment to fifteen minutes.

8. In step 4 of Analysis we ask the student to fit the exponential to only the first 15 minutes of data. This is important, because the fit will sometimes be poorer if all 30 minutes of data are used. The counts during the first 15 minutes are largely due to the barium, while the counts in the last 15 minutes are mostly from non-barium sources. The many noisy points in the tail of the exponential may unduly influence a fit of the entire run.

    You may want to have students investigate this effect, or to try various selections of data during the first 15 minutes (*e.g.*, 2-13 minutes, or 5-15). The resulting value for the lifetime will vary somewhat, giving an indication of the uncertainty of the measurement. Using our data we get variations about 0.05 minutes around the typical value shown here.

9. Note that the calculator and computer versions of the activity use different notation for the fitted equation. Unlike Logger *Pro*, the calculator program DataRad uses seconds as the x-axis time unit, so that the exponential fit parameter must be converted from s^{-1} to min^{-1} ($s^{-1} = 60\ min^{-1}$) to obtain a lifetime in min^{-1}.

SAMPLE RESULTS

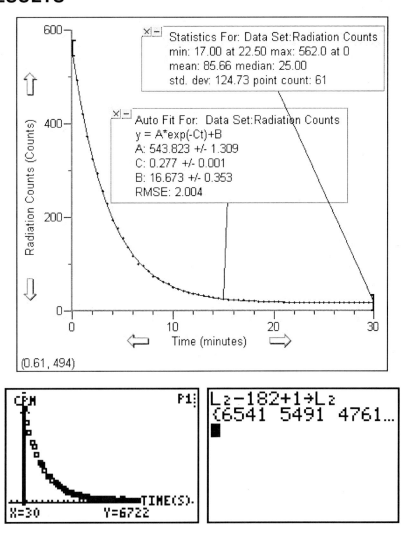

Statistics For: Data Set:Radiation Counts
min: 17.00 at 22.50 max: 562.0 at 0
mean: 85.66 median: 25.00
std. dev: 124.73 point count: 61

Auto Fit For: Data Set:Radiation Counts
$y = A*exp(-Ct)+B$
A: 543.823 +/- 1.309
C: 0.277 +/- 0.001
B: 16.673 +/- 0.353
RMSE: 2.004

(0.61, 494)

Raw data from calculator and background subtraction step.

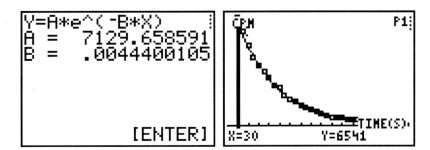

Exponential fit to the first 15 minutes of data after background subtraction.

ANSWERS TO PRELIMINARY QUESTIONS

1. Graph is a decaying exponential. The first few values are 1000, 900, 810... (with integer part of 10% taken each time).

2. Second graph decays more quickly: 1000, 800, 640…

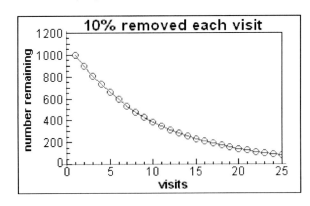

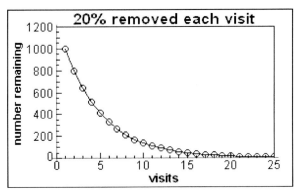

DATA TABLE

(computer)

Average background counts	17
fit parameters for Y = A exp (− C*X) + B	
A	554
B	17
C	0.277
λ (min^{-1})	0.277
$t_{1/2}$ (min)	2.50

(calculator)

fit parameters for Y = A exp (− B*X) + C	
A	7129 cpm
B	0.00444 s^{-1}
C	182 cpm
λ (min^{-1})	0.266
$t_{1/2}$ (min)	2.60

ANSWERS TO ANALYSIS QUESTIONS

1. The count rate decreases in time, falling to less than 10% of the initial value. This is consistent with activity being proportional to the amount of remaining radioactive material, since as material decays, less remains, so the activity must decrease.

2. The three graphs have a similar decreasing shape, although the time-axis scale of the barium data is different from that of the candy graphs. The vertical axes have different units (candy remaining and counts/interval). They are similar because in each case the decay process proceeds at a rate proportional to the remaining candies or radioactive nuclei.

3. We start with the rate equation, and then use the definition of the half-life as the time it takes for the activity to drop to one-half the original value:

$$R = R_0 e^{-\lambda t} \; ; \; \text{at } t = t_{1/2} \; R = \tfrac{1}{2} R_0$$

$$\frac{1}{2} = e^{-\lambda t_{1/2}}$$

$$-\ln 2 = -\lambda t_{1/2}$$

$$t_{1/2} = \frac{\ln 2}{\lambda}$$

4. The experimental half-life of 2.50 min is close to the accepted value of 2.552 s.

5. After 25 minutes, 0.11% of the original barium activity remains. ($e^{-25*0.272} = 0.0011$). Most, but not quite all of the original activity has decayed. The assumption that the counts observed during the last five minutes of data collection are due only to non-barium is reasonable. Possibly a better background estimate could be obtained by waiting a longer time.

ANSWERS TO EXTENSIONS

1. A graph of ln(counts/interval) *vs.* time should be a straight line of negative slope. The slope is $-\lambda$, or the negative of the decay constant. If the background has been subtracted, the graph should be nearly linear. Without background subtraction, the graph will be curved.

2. Results will vary. A collection of lifetime measurements will allow the student to determine a range of values; the extent of that range is a measure of the uncertainty of the measurement. The range of data selected will also influence the measurement, as will the value used for the additive parameter B in the exponential curve fit.

3. Results will depend on the background radiation level. Experiments done at high altitude will experience larger background count rates due to reduced attenuation of cosmic rays by the atmosphere. To measure the background rate, set up Logger *Pro* to count with no source present. Note that the solution obtained from the isogenerator will contain some cesium, raising the count rate further above background from environmental radiation.

Counting Statistics

Radioactive decays follow some curious rules that are a consequence of quantum mechanics. Regardless of when a particular nucleus was created, all nuclei of the same species (Cobalt-60 in this experiment) have exactly the same probability of decay. We might expect that the longer a nucleus has been around, the more likely it is to decay, but that is not what is observed. Even though the *probability* that a given nucleus will decay is fixed, there is no way to predict *when* it will decay. In this sense the decay process is completely random. Despite this randomness, a collection of many identical and independent nuclei will exhibit certain predictable behaviors, such as a consistent average decay rate when measured over a long time.

There are still variations in the average count rate when measured over a shorter time, however. Suppose we collect data on the number of decays during a five-second interval. We count decays for five seconds, and then another five, and so forth. If the average number of counts during each interval is n, then we will find that the standard deviation of the collection of measurements is on average $n^{1/2}$. The standard deviation is a measure of how far away, on average, a measurement is from the mean value. A histogram of the measurements of the number of decays detected each interval will show the characteristic distribution known as the *Poisson distribution*.

When the average number of decays each interval is small, such as one or two, then the Poisson distribution is not symmetric. An asymmetric distribution means that the most common value is different from the average value. If the average number of decays in each time interval is larger, such as more than twenty, the shape of the Poisson distribution approaches the shape of the Normal, or Gaussian, distribution. The Normal distribution is sometimes called the *bell-shaped curve*, although there are other distributions that also look like a bell! The Normal distribution is symmetric, with the average value being identical to the most common value.

In this experiment you will collect data from a source that exhibits an essentially constant decay rate. Because the lifetime of the source is so long, the average decay rate will not change during your experiment. The interval-to-interval count rate will vary, however, but in a way consistent with the Poisson distribution.

OBJECTIVES

- Use a radiation counter to determine the distribution of count rates from a nearly constant-rate source.
- Compare the distribution of experimental nuclear counting data to the Poisson distribution.
- Observe the gradual transition of count distribution from Poisson statistics to Gaussian statistics as the average count rate increases.

MATERIALS

computer	Vernier Radiation Monitor or
Vernier computer interface	Student Radiation Monitor
Logger *Pro*	Cobalt-60 1 µC disc gamma source

PRELIMINARY QUESTIONS

1. Switch your radiation monitor to AUDIO mode, and place it about ten centimeters from your Co-60 source. When the monitor detects a by-product of a decaying Co-60 nucleus (a gamma ray, in this case) it emits a beep and the red LED flashes. Is there a uniform time between beeps? Or does the time vary? From listening to the sequence of beeps, can you predict when the next beep will occur?

2. Now move the source closer to the monitor. Did the average rate of beeps appear to change? If so, how did it change? Is there any more or less uniformity to the time interval between beeps compared to the slower rate?

PROCEDURE

1. Connect the radiation monitor to DIG/SONIC 1 of the interface. Switch the Monitor to the ON (no audio) position.

2. Prepare the computer for data collection by opening the file "04 Statistics" from the *Nuclear Radiation w Computers* folder of Logger *Pro*. Two graphs are displayed: a histogram of the count rate and the counts *vs.* time.

3. Place the radiation monitor on top of or adjacent to the Co-60 source so that the rate of flashing of the red LED is maximized.

4. Click ▶ Collect to begin collecting data. Logger *Pro* will begin counting the number of gamma photons that strike the detector during each one-second count interval. Data collection will continue for just 30 seconds. Do not move the detector or the source for the remainder of data collection.

5. After data collection is complete, the ▶ Collect button will reappear.

6. To study the variation in count rate distributions, you will need to change the length of one time interval so that the average number of counts is first small (about 1) and then larger (30 or so). The count rate from your particular source depends on its age and initial activity, so you will need to first determine the average count rate from your sample. To do this, click on the graph of counts/interval *vs.* time. Then click on the statistics button 📊 to see the average count rate. Enter this value in your data table. Now determine the necessary interval lengths to achieve an average of 1 count per interval and an average of 30 counts per interval. Round these values up to the next 0.05 second. For example, let's say your average count rate was 4.67 counts per one-second interval. To get about one count per interval with the same source, you would use an interval of 4.67^{-1} or 0.21 seconds, rounded to 0.25 seconds. For 30 counts, multiply this by 30, getting 6.45 s after rounding up. Enter these values in your data table.

7. Set the counting interval to the value needed to obtain an average count of approximately one. To do this, select Data Collection from the Experiment menu. Change the number of seconds/sample to the low count rate value in your data table. Take care not to use the samples/second field. Then, change the Experiment Length field so that 200 samples will be collected (*i.e.*, enter a value 200 times the seconds/sample time). Click Done .

8. Click the ▶ Collect button to begin counting. Observe the histogram as data are collected. Is there a regular pattern as to the next count rate that appears? Do the values appear to be clustered around a most-common value? When data collection is complete, the ▶ Collect button will reappear.

9. After data collection is complete, click the Counts *vs.* Time graph to make it active, and then click on the statistics button to calculate statistics for the data. Record the average and standard deviation in your data table. Rescale your graph if needed.

10. Print your screen by selecting Print from the File menu.

11. Set the counting interval to the value needed to obtain an average count of approximately one. To do this, select Data Collection from the Experiment menu. Change the number of seconds/sample to the high count rate value in your data table. Then, change the Experiment Length field so that 200 samples will be collected (*i.e.*, enter a value 200 times the seconds/sample). Click [Done].

12. Click the [I▶ Collect] button to begin counting. Observe the histogram as data are collected. Is there a regular pattern as to the next count rate that appears? Do the values appear to be clustered around a most-common value? When data collection is complete, the [I▶ Collect] button will reappear. Rescale your graph as needed.

13. Click the counts/interval *vs.* time graph to make it active, and then click on the statistics button to calculate statistics for the data. Record the average and standard deviation in your data table.

14. Print your screen by selecting Print from the File menu.

15. The standard deviation is a measure of how far away, on average, a typical measurement (of counts during each interval) is from the average of all the measurements. The interval defined by (average ± one standard deviation) contains most of the measurements. From your average and standard deviation values, determine this interval, rounded to the nearest integer. From the Histogram data table window, determine the fraction of the measurements that fall within the interval.

DATA TABLE

Average count rate (1 s interval)	

	Low count rate (~1/interval)	High count rate (~30/interval)
Interval length (s)		
Average rate (counts/interval)		
Square root (average rate)		
Standard deviation (counts/int)		
Fraction within ± std dev		

ANALYSIS

1. Is your first histogram (with the low average count rate) symmetric? How can you tell? Was that shape consistent with the Normal distribution?

2. Is your second histogram (with the high average count rate) symmetric? How can you tell? Is the symmetry of your data distribution consistent with the Normal distribution?

3. Calculate the square root of the average count rate for your low- and high-count-rate trials. The square root of the number of counts measured in one interval is an estimate of the standard deviation of a set of measurements, when those measurements follow the Poisson distribution. How does the square-root estimate compare to the actual standard deviation of your set of measurements?

4. Use the comparison in the previous question to answer this question: An experiment yields 900 counts in one interval. Predict the standard deviation of a set of 200 additional measurements made under the same conditions.

5. For your high-count-rate data, is the fraction of the measurements that fall within the interval close to two-thirds? The Normal distribution is symmetric and has two-thirds of its values within one standard deviation of the average. Is the distribution of your data consistent with the Normal distribution?

EXTENSIONS

1. Consult a statistics or nuclear physics reference book to learn the mathematical form of the Poisson distribution. Plot a Poisson distribution with the same average and standard deviation as your low-count-rate data on the same graph with those data.

2. Consult a statistics or nuclear physics reference book to learn the mathematical form of the Normal distribution. Plot a Normal distribution with the same average and standard deviation as your high-count-rate data on the same graph with those data.

3. Determine the fraction of your measurements falling with two standard deviations of the average for the high-count-rate measurements. The Normal distribution includes 90% of the measurements within two standard deviations of the average.

4. Determine the fraction of your measurements falling with three standard deviations of the average for the high-count-rate measurements. The Normal distribution includes 99% of the measurements within two standard deviations of the average.

5. Collect additional data at the high count rate. Use intervals with 500 to 1000 counts. Is the histogram different in shape from your earlier data?

Counting Statistics

Radioactive decays follow some curious rules that are a consequence of quantum mechanics. Regardless of when a particular nucleus was created, all nuclei of the same species (Cobalt-60 in this experiment) have exactly the same probability of decay. We might expect that the longer a nucleus has been around, the more likely it is to decay, but that is not what is observed. Even though the *probability* that a given nucleus will decay is fixed, there is no way to predict *when* it will decay. In this sense the decay process is completely random. Despite this randomness, a collection of many identical and independent nuclei will exhibit certain predictable behaviors, such as a consistent average decay rate when measured over a long time

There are still variations in the average count rate when measured over a shorter time, however. Suppose we collect data on the number of decays during a five-second interval. We count decays for five seconds, and then another five, and so forth. If the average number of counts during each interval is n, then we will find that the standard deviation of the collection of measurements is on average $n^{1/2}$. The standard deviation is a measure of how far away, on average, a measurement is from the mean value. A histogram of the measurements of the number of decays detected each interval will show the characteristic distribution known as the *Poisson distribution*.

When the average number of decays each interval is small, such as one or two, then the Poisson distribution is not symmetric. An asymmetric distribution means that the most common value is different from the average value. If the average number of decays in each time interval is larger, such as more than twenty, the shape of the Poisson distribution approaches the shape of the Normal, or Gaussian, distribution. The Normal distribution is sometimes called the *bell-shaped curve*, although there are other distributions that also look like a bell! The Normal distribution is symmetric, with the average value being identical to the most common value.

In this experiment you will collect data from a source that exhibits an essentially constant decay rate. Because the lifetime of the source is so long, the average decay rate will not change during your experiment. The interval-to-interval count rate will vary, however, but in a way consistent with the Poisson distribution.

OBJECTIVES

- Use a radiation counter to determine the distribution of count rates from a nearly constant-rate source.
- Compare the distribution of experimental nuclear counting data to the Poisson distribution.
- Observe the gradual transition of count distribution from Poisson statistics to Gaussian statistics as the average count rate increases.

MATERIALS

TI Graphing Calculator
LabPro or CBL 2
DataRad Calculator Program

Vernier Radiation Monitor or
    Student Radiation Monitor
Cobalt-60 1-µC disc gamma source

PRELIMINARY QUESTIONS

1. Switch your radiation monitor to AUDIO mode, and place it about ten centimeters from your Co-60 source. When the monitor detects a by-product of a decaying Co-60 nucleus (a gamma ray, in this case) it emits a beep and the red LED flashes. Is there a uniform time between beeps? Or does the time vary? From listening to the sequence of beeps, can you predict when the next beep will occur?

2. Now move the source closer to the monitor. Did the average rate of beeps appear to change? If so, how did it change? Is there any more or less uniformity to the time interval between beeps compared to the slower rate?

PROCEDURE

1. Connect the radiation monitor to DIG/SONIC 1 of the LabPro or to DIG/SONIC of the CBL 2. Use the black link cable to connect the TI graphing calculator to the interface. Firmly press in the cable ends. Switch the monitor to the ON (no audio) position.

2. Turn on the calculator and start the DataRad program. Press ⌈CLEAR⌉ to reset the program.

3. Prepare the DataRad program for this experiment.

    a. Select SETUP from the main screen.
    b. Select SET INTERVAL from the SETUP MENU.
    c. Select SET INTERVAL from the INTERVAL SETTINGS menu.
    d. Enter "100" as the count time interval in seconds. Always complete number entries with ⌈ENTER⌉.
    e. Select OK from the INTERVAL SETTINGS menu.
    f. Select SINGLE INTERVAL from the SETUP MENU.

4. Place the radiation monitor on top of or adjacent to the Co-60 source so that the rate of flashing of the red LED is maximized.

    a. Select START from the main screen to start counting.

5. The calculator will begin counting the number of gamma photons that strike the detector during the 100-second count interval. Do not move the detector or the source for the remainder of data collection.

6. After data collection is complete number of counts/interval detected will be shown on the calculator screen. Enter this value in your data table. Press ⌈ENTER⌉ to return to the main screen.

7. To study the variation in count rate distributions, you will need to change the length of one time interval so that the average number of counts is first small (1 or so) and then larger (30 or so). From the count rate measured in the previous step, determine the necessary interval lengths to achieve an average of one count per interval and an average of thirty counts per interval. Round the values up to the next 0.05 s. For example, let's say your average count rate was 530 counts during your 100-second interval. To get about one count per interval with the same source, you would use an interval of $(530/100)^{-1}$ or 0.18 seconds, rounded to 0.20 seconds. For 30 counts, multiply this by 30, getting 6.0 seconds. Enter these rounded values in your data table.

8. Set the counting interval to the value needed to obtain an average count of approximately one.

   a. Select SETUP from the main screen.
   b. Select SET INTERVAL from the SETUP MENU.
   c. Select SET INTERVAL from the INTERVAL SETTINGS screen.
   d. Enter "yy" as the count time interval in seconds, where yy is the count interval you determined for the lower average count rate.
   e. Select OK from the INTERVAL SETTINGS screen.
   f. Select RATE/HISTOGRAM from the SETUP MENU.

9. Select START to begin counting. Observe the count rate values as data are collected. Is there a regular pattern as to the next count rate that appears? Do the values appear to be clustered around a most-common value? Continue collecting data until at least 200 intervals have elapsed. Some time after the 200th interval, press (STO▶) to end data collection.

10. DataRad will display a histogram of the measured count rates. Note that the DataRad program reports the count rate in counts per minute, so these values will be large. Initially there may be a small number of bins in the histogram, which may introduce artificial gaps in the graph. To see the histogram with a different number of bins, press (ENTER) and select YES. Enter a number to see the same data in a histogram with a different number of bins. Print or sketch your histogram.

11. Press (ENTER) and select NO to see the count statistics. Record the average and standard deviation count rates in your main data table.

12. Now set the counting interval to the value needed to obtain an average count of approximately thirty. The larger average count rate will significantly change the shape of the distribution of count rates. Use the same method as you did earlier to increase the interval setting, and select the rate/histogram mode.

13. Select START to begin counting. Observe the histogram as data are collected. Is there a regular pattern as to the next count rate that appears? Do the values appear to be clustered around a most-common value? Continue collecting data until 200 intervals have elapsed. After the 200th interval, press (STO▶) to end data collection.

14. Create your own two-column data table for the histogram data. Label one column "bin max," or the maximum count value in a histogram bin. Label the other "number," or the number of times that measurements of that range occurred. Use the (◄) and (►) keys to read the ten pairs of values from your histogram, recording them in your data table. You will use this table in Analysis. Print or sketch your histogram.

15. Press (ENTER) and select NO to see the count statistics. Record the average and standard deviation in your main data table.

16. The standard deviation is a measure of how far away, on average, a typical measurement (of counts during each interval) is from the average of all the measurements. The interval defined by (average ± one standard deviation) contains most of the measurements. From your average and standard deviation values, determine this interval, rounded to the nearest integer. Then, from your histogram data table, mark the first bin that contains counts corresponding to average − one standard deviation, remembering that the bin labels you recorded were the maximum values. Similarly mark the last bin that contains counts corresponding to average + one standard deviation. Using the total of the number of counts in these bins and those in between, determine the fraction of the measurements that fall within the interval (average ± one standard deviation).

DATA TABLE

Counts/interval (100 s interval)	

	Low count rate (~1/interval)	High count rate (~30/interval)
Interval length (s)		
Average rate (cpm)		
Average counts		
Square root (average counts)		
Standard deviation (cpm)		
Standard deviation (counts)		
Fraction within ± std dev		

ANALYSIS

1. Is your first histogram (with the low average count rate) symmetric? How can you tell? Is that shape consistent with the Normal distribution?

2. Is your second histogram (with the high average count rate) approximately symmetric? How can you tell? Is the symmetry of your data distribution consistent with the Normal distribution?

3. The DataRad program reports count rates in cpm, or counts per minute. Knowing the interval length you used, calculate the number of counts actually detected during an average interval. Enter these values in your data table. Similarly, the program reports the standard deviation of the measurements in cpm. Convert this value to the standard deviation in counts detected during each interval. Enter these values in your data table.

4. Calculate the square root of the average count rate for your low and high count rate trials. The square root of the number of counts measured in one interval is an estimate of the standard deviation of a set of measurements, when those measurements follow the Poisson distribution. How does the square-root estimate compare to the actual standard deviation of your set of measurements?

5. Use the comparison in the previous question to answer this question: An experiment yields 900 counts in one interval. Predict the standard deviation of a set of 200 additional measurements made under the same conditions.

6. Is the fraction of the measurements that fall within the interval close to two-thirds? The Normal distribution is symmetric and has two-thirds of its values within one standard deviation of the average. Is the distribution of your data consistent with the Normal distribution? Remember that you used only ten bins in the histogram, and so your fraction includes count rates not in the desired interval.

EXTENSIONS

1. Use additional bins (45) in your histograms to make more careful determinations of the fraction of measurements that fall within one standard deviation of the average count rate. Use these more detailed histograms in all of the investigations below.

2. Consult a statistics or nuclear physics reference book to learn the mathematical form of the Poisson distribution. Plot a Poisson distribution with the same average and standard deviation as your low-count-rate data on the same graph with those data.

3. Consult a statistics or nuclear physics reference book to learn the mathematical form of the Normal distribution. Plot a Normal distribution with the same average and standard deviation as your high-count-rate data on the same graph with those data.

4. Determine the fraction of your measurements falling with two standard deviations of the average for the high-count-rate measurements. The Normal distribution includes 90% of the measurements within two standard deviations of the average.

5. Determine the fraction of your measurements falling with three standard deviations of the average for the high-count-rate measurements. The Normal distribution includes 99% of the measurements within two standard deviations of the average.

TEACHER INFORMATION

Counting Statistics

1. The experiment calls for counting times that allow for 200 intervals. This is a minimal collection time. If 400 to 500 intervals are used, the resulting distributions will be cleaner and more "like the textbook." If a fresh 1-μC source is used, the 500-interval count will not take excessively long. Use 500 intervals if time allows. Consider collecting data overnight for even better data.

2. You may want to have your students collect low-rate and high-rate data several times to compare the histograms. Particularly if only 200 intervals are used, the histograms will vary from run to run. If more intervals are used, the histograms will vary much less from run to run.

3. It is interesting to watch the histogram "grow" during data collection. You'll see the pattern start as a very rough pattern as the first few bars appear, and then as more and more data are collected the pattern will fill in to approximate an ideal distribution. You may want to have your students observe this during data collection.

4. Sources are available from a number of suppliers:

    *   Spectrum Techniques, 106 Union Valley Road, Oak Ridge, TN 37830, (865) 482-9937, Fax: (865) 483-0473, www.spectrumtechniques.com.
    *   Flinn Scientific, P.O. Box 219, Batavia, IL 60510, (800) 452-1261, Fax: (630) 879-6962, www.flinnsci.com.
    *   Canberra Industries, 800 Research Parkway, Meriden, CT 06450, (800) 243-3955 Fax: (203) 235-1347, www.canberra.com.

SAMPLE RESULTS

Low count rate results with asymmetric histogram:

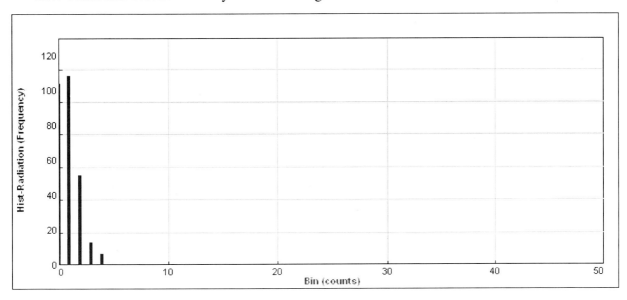

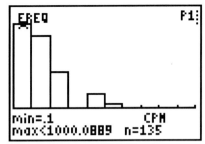 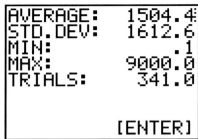

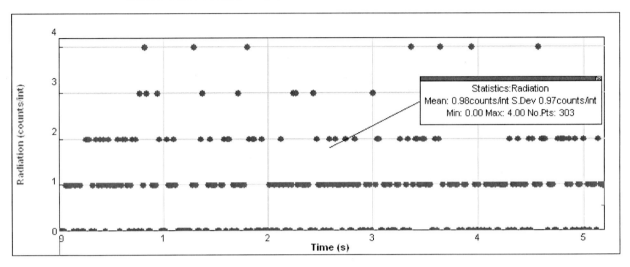

High count rate results with symmetric histogram:

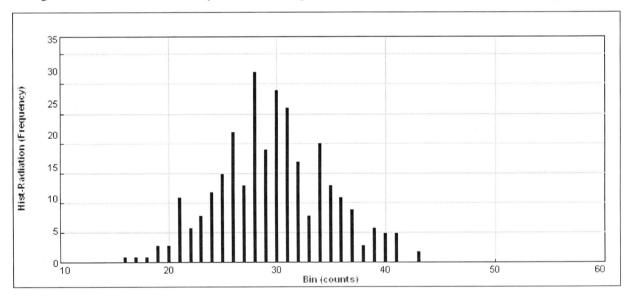

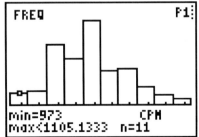 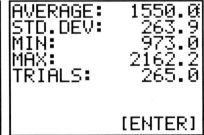

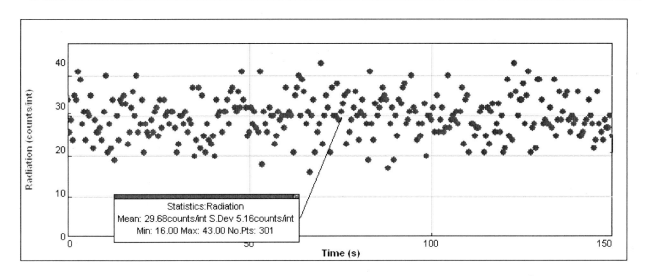

ANSWERS TO PRELIMINARY QUESTIONS

1. The time between beeps (events) is not uniform; sometimes it is long, and sometimes it is very short. There is apparently no way to predict when the next event will occur.

2. There are now more beeps each second, but they are still irregular.

DATA TABLE

(computer data)

Average count rate (1 s interval)	33.7

	Low count rate (~1/interval)	High count rate (~30/interval)
Interval length (s)	0.017	0.9
Average rate (counts/interval)	0.98	28.9
Square root (average rate)	0.99	5.4
Standard Deviation (counts/int)	0.97	5.1
Fraction within ± std dev		70%

(calculator data)

Counts/interval (100 s interval)	2710

	Low count rate (~1/interval)	High count rate (~30/interval)
Interval length (s)	0.04	1.11
Average rate (cpm)	1504	1550
Average counts	1.00	28.7
Square root (average counts)	1.00	5.3
Standard deviation (cpm)	1612	263.9
Standard deviation (counts)	1.07	4.9
Fraction within ± std dev		80 %

Bin max	number
1105	11
1237	12
1369	49
1501	38
1633	69
1765	28
1897	30
2030	15
2162	8
2294	5

214/265 or 80% of the measurements are within one standard deviation of the average.

ANSWERS TO ANALYSIS QUESTIONS

1. No, the low-count rate histogram is not symmetric. This is apparent from the peak that is left of center on the distribution. The asymmetric shape is different from the Normal distribution, so these data are not distributed like the Normal distribution.

2. The second, high-rate, histogram appears symmetric since the peak is in the middle. This shape is qualitatively similar to the Normal distribution.

3. (4 for calculator) The square root estimates are very close to the actual standard deviations. This is consistent with the count data following the Poisson distribution.

4. (5 for calculator) The estimated standard deviation of a set of measurements with a 900-count average would be $900^{0.5}$, or 30. That 200 (or 2000) measurements are to be made is not relevant.

5. (6 for calculator) The fraction of measurements within one standard deviation of the average is 70%, which is very similar to the expected two-thirds of values within that range for the Normal distribution. The calculator histogram bins are broader than are the computer bins, so the higher 80% fraction is due to over counting in the broader bins.

ANSWERS TO EXTENSIONS

1. Below is a histogram of low-count rate data of average rate three (solid bars) with Poisson distribution (hatched bars) of same average and area. This graph was created using a spreadsheet (Microsoft Excel 97). The distribution of the experimental data and the Poisson distribution are very similar; both are asymmetric. The expression used to calculate the unit-area Poisson distribution, where x the counts/interval and μ the average, is

$$P(x;\mu) = \frac{\mu^x}{x!}e^{-\mu}$$

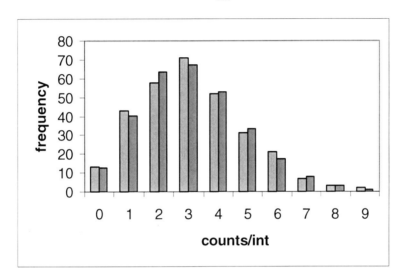

2. Below is a histogram of high-count rate data (solid bars) with Poisson distribution (hatched bars) of same average and area. The distribution of the experimental data and the Poisson distribution are very similar; both are nearly symmetric as expected for data with an average count rate near thirty.

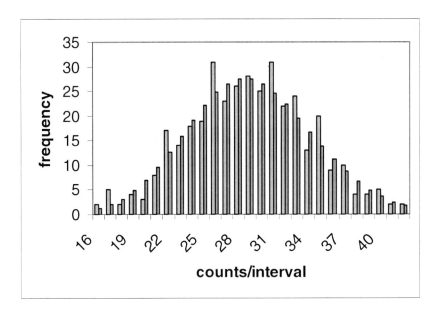

3. Approximately 90% of the measurements fall within an interval two standard deviations on either side of the average. This is consistent with the Normal distribution.

4. Essentially all of the measurements fall within an interval three standard deviations on either side of the average. This is consistent with the Normal distribution, where 99.7% of the measurements fall within this range.

Background Radiation Sources

When a Geiger counter is operated it will usually record an event every few seconds, even if no obvious radioactive source is placed nearby. Where do these counts come from?

Two significant sources are cosmic rays and radon decay products. Cosmic rays, as the name suggests, are fast-moving particles from space that enter the Earth's atmosphere, along with their decay products. Since the atmosphere absorbs some of these particles, the rate of detection of cosmic rays increases with increasing altitude. If you were to take a Geiger counter on a cross-country jet flight, you would observe a marked increase in count rate while at high altitude.

Another radiation source comes not from above us, but from below. The Earth's crust contains, among other radioactive elements, uranium-238 (^{238}U). ^{238}U has a long half-life, but its decay products do not. One of these products is radon gas, or ^{222}Rn. As a result of the long uranium half-life, there is a nearly steady production of radon, which itself decays with a short half-life of 3.8 days. Since radon is a gas, it diffuses out of the soil into the air, and can collect in low enclosed areas such as basements. Radon decays to a series of species including polonium, lead, and bismuth. These decay products precipitate out of the air onto dust particles since they are solids, unlike gaseous radon.

The decay products are also electrically charged, so that it is relatively easy to collect them on a charged surface for analysis. One simple way to create a charged surface is to rub a balloon with fur or hair—you have probably done this to stick balloons to the wall using their static charge. If you allow the balloon to sit undisturbed for 45 minutes or so, it will collect a fresh set of decay products. The beta and gamma ray emissions as these products themselves decay can be detected using a Geiger counter.

The ^{238}U decay sequence relevant to this experiment is ^{222}Rn (3.8 d) $\rightarrow$ ^{218}Po (3.1 min) $\rightarrow$ ^{214}Pb (27 min) $\rightarrow$ ^{214}Bi (20 min) $\rightarrow$ ^{214}Po (164 µs) $\rightarrow$... The times in parenthesis are the half-lives of each species. Since your Geiger counter is unable to distinguish between the emissions from each of these decays, you would only be able to measure a composite effect of some of the components. Nevertheless, if you observe a time-dependent count rate from your balloon, you will have evidence that there must be a continuous re-supply of the parent radon to the environment.

OBJECTIVES

- Concentrate naturally occurring radioactive substances using a charged balloon.
- Use a radiation counter to detect emissions from naturally occurring radioactive substances.
- Determine the effective lifetime of the collection of radon decay products.

MATERIALS

computer
Vernier computer interface
Logger *Pro*

Vernier Radiation Monitor or
  Student Radiation Monitor
toy balloon, hair, or fur for charging
string, 2 m

PRELIMINARY QUESTION

1. Turn on the radiation monitor to the audio mode, so that it beeps when radiation is detected. Can you detect any radiation in your laboratory? Do you have any way to determine what the radiation is coming from?

PROCEDURE

1. Blow up your balloon so that it is firm. Tie the string to the balloon so you can suspend it in mid-air. Rub the entire surface of the balloon vigorously on the fur for about a minute to give it a static charge. Test the charge by picking up small bits of paper with the balloon by bringing it near the paper. The paper should jump to the balloon's surface and stick there.

2. Suspend the charged balloon away from other objects and where it will not be disturbed. An enclosed basement room is best. Be sure the balloon is at least 2 meters from the radiation monitor. Let the balloon remain in position for 45 minutes.

3. Connect the radiation monitor to DIG/SONIC 1 of the computer interface. Switch the monitor to the on (not audio) mode.

4. Prepare the computer for data collection by opening "05 Background" from the *Nuclear Radiation w Computers* experiment files of Logger *Pro*. One graph is displayed: counts *vs.* time. The vertical axis is scaled from 0 to 1000 counts/interval. The horizontal axis is distance scaled from 0 to 200 minutes.

5. While you are waiting for the balloon to collect its radioactive dust, determine the average background count rate from cosmic rays and unconcentrated dust. To do this, click the ▮Collect button and allow the computer to count events. Every five minutes a new point will be added to the graph.

6. When the balloon still has five minutes remaining in its dust collection time, click the ■Stop button of Logger *Pro*. To determine the average background count rate, click once on the graph, and then click the statistics button. Record the average number of counts for each 5-minute interval in your data table.

7. Once the balloon has been in position for 45 minutes, take it down and deflate it. Taking care not to rub the collected dust from the surface, roll it into a small cylinder. Place the roll as close as possible to the window of the radiation monitor.

8. Click the ▮Collect button to begin data collection. Logger *Pro* may ask you what you want to do; click Erase and Continue to start data collection. Wait 200 minutes for Logger *Pro* to complete data collection.

DATA TABLE

Average background count rate in 5 minutes	
fit parameters for Y = A exp (– C*X) + B	
A	
B	
C	
λ (min^{-1})	
$t_{1/2}$ (min)	

ANALYSIS

1. Inspect your graph. Is the count rate from the balloon-concentrated dust greater than the average background rate you observed? Is the difference significant? Does the count rate decrease with time? (If the initial count rate is not higher than background, the balloon may not have collected sufficient radioactive dust for the following analysis to be meaningful.)

2. Fit an exponential function to your data. To do this, click the curve fit button [☑]. Select Natural Exponent from the equation list, and then click [Try Fit]. A best-fit curve will be displayed on the graph. If your data follow the exponential relationship, the curve should closely match the data. You can force the fit to use the average background count rate you determined without the balloon by entering your measured rate in the B value field, and again click [Try Fit]. When you are satisfied with the fit, click [OK].

3. Print or sketch your graph.

4. Record the fit parameters A, B, and C in your data table.

5. From the fit parameters, determine the decay constant λ and the half-life $t_{1/2}$. Is it necessary to correct for background counts from cosmic rays? Note that there is an additive constant in the fitted equation Y = A exp (– C*X) + B. How does the additive constant compare to the background count rate you measured?

6. Is your value of $t_{1/2}$ consistent with any one half-life of the radon decay products?

7. What fraction of the initial activity of your sample would remain after five hours, if you were to continue the experiment for that length of time?

8. Cosmic rays arrive at the Earth's surface at a roughly constant rate. Do you have evidence of the presence of a non-cosmic ray source of radiation in your laboratory? Explain.

9. If radon gas is the source of any time-dependent count rates you observed, must there be a continuous source of fresh radon gas to the environment?

EXTENSIONS

1. Is there a variation in the background radiation count rate in different places in your school? List several reasons why the rate might vary, or why it might be the same.

2. From the decay sequence given in the introduction, determine the type of nuclear decay (alpha, beta, or gamma) in each step.

3. Rather than allowing Logger *Pro* to determine the background count rate with the additive term of the exponential curve fit, correct the experimental count rates by subtracting the average background count rate you measured earlier in the experiment. Plot a new graph of corrected count rate *vs.* time, and fit a new exponential function to the data. Now that you have corrected for background counts, what should happen to the fitted value of the additive constant B in $Y = A \exp(-C*X) + B$?

4. Using the background-corrected data created in the previous extension, create a plot of the natural log of the corrected count rates *vs.* time. What is the significance of the slope of a line fitted to the data?

5. Why would you expect a graph of count rate *vs.* time, such as collected for this experiment, to *not* be a simple exponential function?

Background Radiation Sources

When a Geiger counter is operated it will usually record an event every few seconds, even if no obvious radioactive source is placed nearby. Where do these counts come from?

Two significant sources are cosmic rays and radon decay products. Cosmic rays, as the name suggests, are fast-moving particles from space that enter the Earth's atmosphere, along with their decay products. Since the atmosphere absorbs some of these particles, the rate of detection of cosmic rays increases with increasing altitude. If you were to take a Geiger counter on a cross-country jet flight, you would observe a marked increase in count rate while at high altitude.

Another radiation source comes not from above us, but from below. The Earth's crust contains, among other radioactive elements, uranium-238 (^{238}U). ^{238}U has a long half-life, but its decay products do not. One of these products is radon gas, or ^{222}Rn. As a result of the long uranium half-life, there is a nearly steady production of radon, which itself decays with a short half-life of 3.8 days. Since radon is a gas, it diffuses out of the soil into the air, and can collect in low enclosed areas such as basements. Radon decays to a series of species including polonium, lead, and bismuth. These decay products precipitate out of the air onto dust particles since they are solids, unlike gaseous radon.

The decay products are also electrically charged, so that it is relatively easy to collect them on a charged surface for analysis. One simple way to create a charged surface is to rub a balloon with fur or hair—you have probably done this to stick balloons to the wall using their static charge. If you allow the balloon to sit undisturbed for 45 minutes or so, it will collect a fresh set of decay products. The beta and gamma ray emissions as these products themselves decay can be detected using a Geiger counter.

The ^{238}U decay sequence relevant to this experiment is ^{222}Rn (3.8 d) $\rightarrow$ ^{218}Po (3.1 min) $\rightarrow$ ^{214}Pb (27 min) $\rightarrow$ ^{214}Bi (20 min) $\rightarrow$ ^{214}Po (164 μs) $\rightarrow$... The times in parenthesis are the half-lives of each species. Since your Geiger counter is unable to distinguish between the emissions from each of these decays, you would only be able to measure a composite effect of some of the components. Nevertheless, if you observe a time-dependent count rate from your balloon, you will have evidence that there must be a continuous re-supply of the parent radon to the environment.

OBJECTIVES

- Concentrate naturally occurring radioactive substances using a charged balloon.
- Use a radiation counter to detect emissions from naturally occurring radioactive substances.
- Determine the effective lifetime of the collection of radon decay products.

MATERIALS

TI Graphing Calculator
LabPro or CBL 2
DataRad Calculator Program

Vernier Radiation Monitor or
   Student Radiation Monitor
toy balloon, hair, or fur for charging
string, 2 m

PRELIMINARY QUESTIONS

1. Turn on the radiation monitor to the audio mode, so that it beeps when radiation is detected. Can you detect any radiation in your laboratory? Do you have any way to determine what the radiation is coming from?

PROCEDURE

1. Blow up your balloon so that it is firm. Tie the string to the balloon so you can suspend it in mid-air. Rub it vigorously on the fur for about a minute to give it a static charge. Test the charge by picking up small bits of paper with the balloon by bringing it near the paper. The paper should jump to the balloon's surface and stick there.

2. Suspend the charged balloon away from other objects and where it will not be disturbed. Be sure the balloon is at least 2 meters from the radiation monitor. Let the balloon remain in position for 45 minutes.

3. Connect the radiation monitor to DIG/SONIC 1 of the LabPro or to DIG/SONIC of the CBL 2. Use the black link cable to connect the TI graphing calculator to the interface. Firmly press in the cable ends.

4. Turn on the calculator and start the DataRad program. Press CLEAR to reset the program.

5. Prepare the DataRad program for this experiment.

   a. Select SETUP from the main screen.
   b. Select SET INTERVAL from the SETUP MENU.
   c. Select SET INTERVAL from the INTERVAL SETTINGS screen.
   d. Enter "300" as the count time interval in seconds. Always complete number entries with ENTER.
   e. Select OK from the INTERVAL SETTINGS screen.
   f. Select BACKGROUND CORRECTION from the SETUP MENU.
   g. Select PERFORM NOW from the BACKGROUND CORRECTION screen.
   h. Enter "5" as the number of intervals for background correction.
   i. Confirm that no radiation sources are near the radiation monitor, and press ENTER to start counting.
   j. After counting is complete, the background correction rate will be displayed. Record the average background count rate (in counts per minute, or cpm) in your data table.
   k. Press ENTER to return to the SETUP MENU.

6. Your balloon should have a few minutes left in its dust collection time. Prepare the calculator to collect count data as a function of time.

   a. Select TIME GRAPH from the SETUP MENU.
   b. Select CHANGE TIME SETTINGS from the TIME GRAPH SETTINGS screen.
   c. Enter "300" as the count time interval in seconds.
   d. Enter "40" as the number of samples. This setting will give you a 300*40=12000 second (200 minute) data collection time.
   e. Select OK from the TIME GRAPH SETTINGS screen to return to the main screen.

7. Once the balloon has been in position for 45 minutes, take it down and deflate it. Taking care not to rub the collected dust from the surface, roll it into a small cylinder. Place the roll as close as possible to the window of the radiation monitor.

8. Select START to begin data collection. Wait 200 minutes for data collection to complete.

DATA TABLE

Average background count rate in 5 minutes (cpm)	
fit parameters for Y = A exp (− B*X)	
A	
B	
λ (min^{-1})	
$t_{1/2}$ (min)	

ANALYSIS

1. Inspect your graph. Is the count rate from the balloon-concentrated dust greater than the average background rate you observed? Is the difference significant? Does the count rate decrease with time? (If the initial count rate is not higher than background, the balloon may not have collected sufficient radioactive dust for the following analysis to be meaningful.)

2. Fit an exponential function to your data.

   a. Select ANALYZE.
   b. Select EXPONENT CURVE FIT from the CURVE FITS screen.
   c. Record the fit parameters in your data table.
   d. Press (ENTER) to see the fitted curve with your data. If your data follow the exponential relationship, the curve should closely match the data.

3. Print or sketch your graph.

4. From the fit parameters, determine the decay constant λ and the half-life $t_{1/2}$.

5. Is your value of $t_{1/2}$ consistent with any one half-life of the radon decay products?

6. What fraction of the initial activity of your sample would remain after five hours, if you were to continue the experiment for that length of time?

7. Cosmic rays arrive at the Earth's surface at a roughly constant rate. Do you have evidence of the presence of a non-cosmic ray source of radiation in your laboratory? Explain.

8. If radon gas is the source of any time-dependent count rates you observed, must there be a continuous source of fresh radon gas to the environment?

EXTENSIONS

1. Is there a variation in the background radiation count rate in different places in your school? List several reasons why the rate might vary, or why it might be the same.

2. From the decay sequence given in the introduction, determine the type of nuclear decay (alpha, beta, or gamma) in each step.

3. Create a plot of the natural log of the corrected count rates *vs.* time. What is the significance of the slope of a line fitted to the data?

4. Why would you expect a graph of count rate *vs.* time, such as collected for this experiment, to *not* be a simple exponential function?

Background Radiation Sources

1. This experiment is based on ideas found in several papers: "Radioactiveball," by James Cowie, Jr., and Thomas A. Walkiewicz, *The Physics Teacher* **30**, Jan, 1992, 16, and "The Hot Balloon (Not Air)," by Thomas A. Walkiewicz, *The Physics Teacher* **33**, Sept., 1995, 344.

2. Since the radon decay products are produced as ions, we collect them electrostatically. The air must be dry for effective collection. If you cannot charge the balloon so it will strongly attract hair and dust, it is too humid and you will collect little if any decay products. You may want to reserve this experiment for a dry winter day. If the count rate from the concentrated dust is not significantly different from background, then you have not collected a sufficient quantity of airborne radioactive dust. This could be due to high humidity, or it could be due to a low local radon concentration. It is possible that you will get little or no counts above background. Below is a map from the U.S. Geological Survey showing the potential for radon presence across the United States. Note that just because a region has the potential to be a high-radon area does not mean that a given building will necessarily have a high radon level. Source: http://energy.cr.usgs.gov/radon/rnus.html

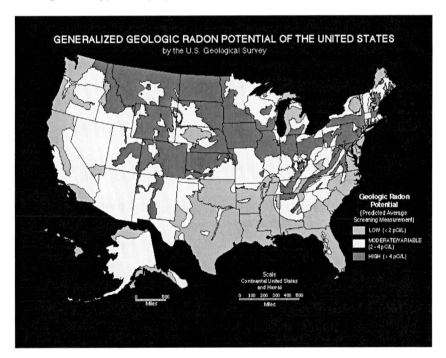

3. It is important to allow the balloon to remain undisturbed for 45 minutes. This time allows the decay products captured by the balloon to roughly reach secular equilibrium. You will need to allow time for this in class, in addition to the 200-minute data collection period. If you only have one class period available, have your students charge and place the balloon immediately, and then allow data collection to continue after the students have left. Analysis can then be performed at another time.

4. Some televisions and computer monitors acquire a static charge, and so they collect dust rapidly. If you have such a monitor you could use an alternative dust collection method.

Clean the monitor thoroughly, and then let it operate undisturbed for 45 minutes. Using a clean piece of tissue or lens paper, wipe the screen clean with a small area of the paper. Use this dust sample for the same experiment as described in the student activity. Do not use the dust that has been on the screen for a long period, as it will represent a different equilibrium population of radon progeny with a longer half-life.

5. Students sometimes confuse the decay constant parameter λ with the half-life $t_{1/2}$. The decay constant λ is larger for more rapidly decaying elements and has dimensions of time^{-1}, while the half-life has dimensions of time, and is smaller for more rapidly decaying elements. The decay constant λ is equal to the fit parameter C. The two parameters can be related in the following manner. After one half-life has elapsed, half of the radioactive nuclei have decayed, and so the activity is also cut in half. From the rate equation we can relate the decay constant to the half life:

$$R = R_0 e^{-\lambda t} \ ; \text{at } t = t_{1/2} \ \ R = \tfrac{1}{2} R_0$$

$$\frac{1}{2} = e^{-\lambda t_{1/2}}$$

$$-\ln 2 = -\lambda t_{1/2}$$

$$t_{1/2} = \frac{\ln 2}{\lambda}$$

The experiment file supplied with this activity uses time units of minutes, so the decay constant and half-life will most naturally be calculated in min^{-1} and min, respectively.

6. Since the dust sample collected by the balloon is not a single nuclear species decaying to a stable state, the resulting decay curve is not a single exponential, but a sum of several related exponential functions. The effective half-life measured here does not correspond to any one decay element. The value numerical value of the effective half-life will depend on several factors, including the length of time the balloon is allowed to collect dust and the time between deflating the balloon and beginning data collection. For a 45-minute data collection and a 3-minute delay before starting data collection, a typical half-life is about 40 minutes.

7. For counting measurements you can estimate the standard deviation of a set of measurements using the square root of the number of counts. For example, if you measured a rate of 100 counts in five minutes, the standard deviation would be 10 counts in five minutes. The standard deviation can be used as a measure of the uncertainty of a count rate measurement.

8. A longer version of this experiment can be found in the *Workshop Physics Activity Guide* Module 3, Priscilla W. Laws, John Wiley & Sons, ©1996, unit "Radioactivity and Radon." A spreadsheet model of the radon decay product series can explain the observed half-life.

9. Note that the calculator and computer versions of the activity use different notation for the fitted equation. The calculator version also uses seconds as the x-axis time unit, so that the exponential fit parameter must be converted from s^{-1} to min^{-1} (s^{-1} = 60 min^{-1}) to obtain a lifetime in min^{-1}.

SAMPLE RESULTS

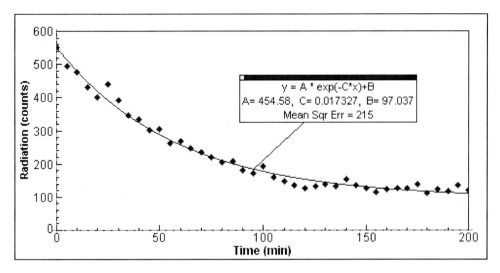

ANSWERS TO PRELIMINARY QUESTIONS

1. Yes, there is radiation in the laboratory since the radiation monitor beeps with no obvious source nearby. Since no source is apparent, there is no clear way as yet to determine the source of the radiation.

DATA TABLE

Average background count rate in 5 minutes	103
fit parameters for Y = A exp (– C*X) + B	
A	454
B	97
C	0.0173
λ (min^{-1})	0.0173
$t_{1/2}$ (min)	40

ANSWERS TO ANALYSIS QUESTIONS

1. Yes, the initial number of counts in a 5-minute interval is greater than the average background rate by a factor of (five), which is a significant difference in terms of the standard deviation of the count measurement. The count rate for the balloon-concentrated dust does decrease with time.

5. Since the fitted equation has an additive constant B, there is no need to correct for the cosmic-ray background. The coefficients of the exponential will reflect only the changing component of the count rates. The additive constant B from the curve fit is nearly the same as the average background count rate without the balloon.

6. The measured half-life (40 minutes) is significantly longer than any of the individual half-lives of the radon decay products. As a result we can conclude that we are observing a combination of decays from several products.

7. Given a 40-minute half-life, we have $2^{-5*60/40} = 0.055$, or about 6% of the initial activity remains after five hours. This does not include the constant-rate background counts from sources other than the balloon.

8. Since the count rate from the balloon-concentrated dust decreased in time, there must be some radioactive source in the environment other than the constant-rate cosmic rays. Radon is a likely source since, as a gas, it can easily spread from the soil.

9. If we assume that the radioactive dust collected by the balloon consisted of radon decay products, then there must be a continuous introduction of radon into the environment. If this were not the case, then the radon decay products would have long ago decayed and we would not detect their radiation.

ANSWERS TO EXTENSIONS

1. Answers will vary, but unless there is a location with very high radon concentration, students will generally find that the background count rate is independent of location within a single community. On the other hand, if radon concentrations vary or if the building acts as a significant cosmic ray shield, then the background will vary with location.

2. ^{222}Rn (3.8 d) $\rightarrow$ ^{218}Po (3.1 min) $\rightarrow$ ^{214}Pb (27 min) $\rightarrow$ ^{214}Bi (20 min) $\rightarrow$ ^{214}Po (164 μs) $\rightarrow$...
Radon decays by alpha emission, followed by alpha, beta, and then beta emission.

3. The additive constant should be small in comparison to its original value before background subtraction.

4. A log plot of a single exponential function would be a straight line, with the slope equal to the decay constant. Since these data roughly follow a straight line we can assign an effective decay constant or half-life to the dust sample collected by the balloon. The points at longer times will show significant scatter.

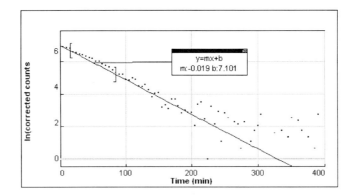

5. Since the dust sample collected by the balloon is not a single nuclear species decaying to a stable state, the resulting decay curve is not a single exponential, but a sum of several related exponential functions. The effective half-life measured here does not correspond to any one decay element. The numerical value of the effective half-life will depend on several factors, including the length of time the balloon is allowed to collect dust and the time between deflating the balloon and beginning data collection.

Radiation Shielding

Alpha, beta, gamma, and X-rays can pass through matter, but can also be absorbed or scattered in varying degrees depending on the material and on the type and energy of the radiation. Medical X-ray images are possible because bones absorb X-rays more so than do soft tissues. Strongly radioactive sources are often stored in heavy lead boxes to shield the local environment from the radiation.

Some materials absorb beta rays. A sheet of common cardboard will absorb some of the betas, but will allow most to pass through. You can measure this absorption by fixing a beta source and a radiation monitor so their positions do not change, and then inserting layers of cardboard between them.

When an absorber is in the path of beta rays, it will allow a certain fraction f to pass through. The fraction f depends on the density and thickness of the absorber, but will be a constant for identical absorbers and fixed beta ray energy. If the number of counts detected in a count interval is N_0 when no absorber is in place, then the counts N with the absorber is $N = f N_0$. In the preliminary questions, you will develop a more general expression for additional layers of cardboard absorbers, and then test it against real data.

In this experiment you will use a small source of beta radiation. Beta rays are high-energy electrons. *Follow all local procedures for handling radioactive materials.*

OBJECTIVES

- Create a model for the absorption of radiation by matter.
- Use a radiation counter to study how the radiation emitted by a beta source is absorbed by cardboard.
- Test the model against experimental data to determine its validity.

MATERIALS

computer	Vernier Radiation Monitor or
Vernier computer interface	Student Radiation Monitor
Logger *Pro*	Strontium-90 0.1µC source taped to
Ten 10 cm × 10 cm identical cardboard squares	small support
	adhesive tape

PRELIMINARY PROCEDURE AND QUESTIONS

1. Place your Sr-90 source on a table. Turn on the radiation monitor to the audio mode, so that it beeps when radiation is detected. By holding the monitor near the source, determine the most sensitive place on the detector.

2. Attach the source disc to a support using adhesive tape so that the source held at the same height as the Geiger tube in the radiation monitor. Do not cover the source with tape. Place the source so it is about eight centimeters from the most sensitive place on the monitor, so that there is room to place all ten layers of cardboard between the source and the monitor. It

is essential that neither the source nor the monitor move during data collection.

With only air between the source and the monitor, listen to the beep rate for a short while. Now place five layers of cardboard between the source and the monitor, taking care not to move either one. Listen again, and determine if the beep rate is larger, smaller, or unchanged. Now add five more layers of cardboard, again not moving the source or monitor. Listen, and determine the change of the beep rate, if any. Does the cardboard seem to shield the monitor from the beta radiation?

3. Based on your observations, sketch a qualitative graph of the beep rate *vs.* number of layers of shielding.

4. In the introduction we used the expression $N = f N_0$ to describe the transmission of betas by one layer of cardboard. Assuming this model, how many counts would be detected if you added a second layer of cardboard, identical to the first, which also transmitted a fraction f? For example, if the first layer transmitted 90% of the radiation, then the second would transmit 90% of that transmitted by the first. The overall transmission would then be $0.90 \times 0.90 = 0.81 = 81\%$ of the no-shielding number of counts. In the data table, write a general expression for the number of counts N detected for any number x of identical layers, each of which transmits a fraction f of the incident radiation. Use N_0 as the counts detected when no shielding layers are used. You have just developed a model for the transmission of radiation through matter. Next you will test your model against experimental data.

5. Is your model consistent with your qualitative graph you sketched based on initial observations? Remember than f is a number less than one. Add the model function to your sketch without worrying about the vertical scale.

PROCEDURE

1. Connect the radiation monitor to DIG/SONIC 1 of the computer interface.

2. Prepare the computer for data collection by opening "06 Shielding" from the *Nuclear Radiation w Computers* experiment files of Logger *Pro*. One graph is displayed: counts *vs.* layers. The vertical axis is scaled from 0 to 1000 counts/interval. The horizontal axis is distance scaled from 0 to 10 layers.

3. Confirm that the source and monitor are positioned so they will not move, and so that there is enough space between them for ten layers of cardboard. Remove all cardboard from between the source and monitor.

4. Click ▶ Collect to begin collecting data. Logger *Pro* will begin counting the number of beta particles that strike the detector during each 50-second count interval.

5. After at least 50 seconds have elapsed, click the Keep button. In the entry field that appears, enter **0**, which is the number of layers of cardboard. Complete your entry by pressing enter on the keyboard. Data collection will now pause for you to adjust the apparatus.

6. Insert one layer of cardboard between the source and detector. Be sure that the cardboard completely covers the source's "view" of the Geiger tube in the detector. Click Continue to collect more data, and wait 50 seconds.

7. Click Keep, and enter the new number of layers, **1**.

8. In the same way as before, add a layer of cardboard without moving the source or monitor, wait 50 seconds, and click ⌈ Keep ⌉. Enter the number of layers of cardboard. Click ⌈ Continue ⌉ to resume data collection. Repeat this process until you have completed data collection for ten layers.

9. Click ⌈Stop Collection⌉ instead of ⌈ Continue ⌉ to end data collection.

DATA TABLE

Model equation	
Fitted equation with parameters	

ANALYSIS

1. Inspect your graph. Does the count rate appear to follow your model?

2. Fit an appropriate function to your data. To choose a function, look for one that has the same mathematical form as your model. To see the functions available, single-click on the graph. Click the curve-fit button ⌈⌣⌉. (Hint: Which fit functions have an x, the horizontal axis variable, in the *same* special location as in your model equation?) Select an equation from the equation list, and then click ⌈ Try Fit ⌉. A best-fit curve will be displayed on the graph. If your data follow the selected relationship, the curve should closely match the data. If the curve does not match well, try a different fit and click ⌈ Try Fit ⌉ again. When you are satisfied with the fit, click ⌈ OK ⌉.

3. Print or sketch your graph. Record the fitted equation and parameters in your data table.

4. From the evidence presented in your graph, does the transmission of beta radiation through cardboard match that predicted by your model?

5. From the parameters of your fitted equation, determine the fraction f of beta rays transmitted, on average, by one layer of cardboard. Do not use your raw data to calculate the fraction, but instead use the better information from your fitted equation. Hint: Remember that $A^{(Bx)} = (A^B)^x$.

EXTENSIONS

1. Use a longer counting interval so that you collect at least 2000 counts when no absorbing cardboard is in place. Is the agreement with the model any different? Try a much shorter count interval. How is the resulting graph different? Why?

2. Cosmic rays continue to strike the detector regardless of the absorbing cardboard. Measure the average background counts in one count interval, and correct your data for background radiation. Repeat the analysis.

3. Try other absorbers; for example, common household aluminum foil can be used in place of cardboard. You will need to experiment with the appropriate number of layers to use. You may want to add more than one (or five) layers at a time.

Radiation Shielding

Alpha, beta, gamma, and X-rays can pass through matter, but can also be absorbed or scattered in varying degrees depending on the material and on the type and energy of the radiation. Medical X-ray images are possible because bones absorb X-rays more so than do soft tissues. Strongly radioactive sources are often stored in heavy lead boxes to shield the local environment from the radiation.

Some materials absorb beta rays. A sheet of common cardboard will absorb some of the betas, but will allow most to pass through. You can measure this absorption by fixing a beta source and a radiation monitor so their positions do not change, and then inserting layers of cardboard between them.

When an absorber is in the path of beta rays, it will allow a certain fraction f to pass through. The fraction f depends on the density and thickness of the absorber, but will be a constant for identical absorbers and fixed beta ray energy. If the number of counts detected in a count interval is N_0 when no absorber is in place, then the counts N with the absorber is $N = f N_0$. In the preliminary questions, you will develop a more general expression for additional layers of cardboard absorbers, and then test it against real data.

In this experiment you will use a small source of beta radiation. Beta rays are high-energy electrons. *Follow all local procedures for handling radioactive materials.*

OBJECTIVES

- Create a model for the absorption of radiation by matter.
- Use a radiation counter to study how the radiation emitted by a beta source is absorbed by cardboard.
- Test the model against experimental data to determine its validity.

MATERIALS

TI Graphing Calculator
LabPro or CBL 2
DataRad calculator program
Ten 10 cm × 10 cm identical cardboard squares

Vernier Radiation Monitor or
   Student Radiation Monitor
Strontium-90 0.1μC source taped to
   small support
adhesive tape

PRELIMINARY PROCEDURE AND QUESTIONS

1. Place your Sr-90 source on a table. Turn on the radiation monitor to the audio mode, so that it beeps when radiation is detected. By holding the monitor near the source, determine the most sensitive place on the detector.

2. Attach the source disc to a support using adhesive tape so that the source held at the same height as the Geiger tube in the radiation monitor. Do not cover the source with tape. Place the source so it is about eight centimeters from the most sensitive place on the monitor, so that there is room to place all ten layers of cardboard between the source and the monitor. It

is essential that neither the source nor the monitor move during data collection.

With only air between the source and the monitor, listen to the beep rate for a short while. Now place five layers of cardboard between the source and the monitor, taking care not to move either one. Listen again, and determine if the beep rate is larger, smaller, or unchanged. Now add five more layers of cardboard, again not moving the source or monitor. Listen, and determine the change of the beep rate, if any. Does the cardboard seem to shield the monitor from the beta radiation?

3. Based on your observations, sketch a qualitative graph of the beep rate *vs.* layers of shielding.

4. In the introduction we used the expression $N = f N_0$ to describe the transmission of betas by one layer of cardboard. Assuming this model, how many counts would be detected if you added a second layer of cardboard, identical to the first, which also transmitted a fraction f? For example, if the first layer transmitted 90% of the radiation, then the second would transmit 90% of that transmitted by the first. The overall transmission would then be $0.90 \times 0.90 = 0.81 = 81\%$ of the no-shielding number of counts. In the data table, write a general expression for the number of counts N detected for any number x of identical layers, each of which transmits a fraction f of the incident radiation. Use N_0 as the counts detected when no shielding layers are used. You have just developed a model for the transmission of radiation through matter. Next you will test your model against experimental data.

5. Is your model consistent with your qualitative graph you sketched based on initial observations? Remember than f is a number less than one. Add the model function to your sketch without worrying about the vertical scale.

PROCEDURE

1. Connect the radiation monitor to DIG/SONIC 1 of the LabPro or to DIG/SONIC of the CBL 2. Use the black link cable to connect the TI graphing calculator to the interface. Firmly press in the cable ends.

2. Turn on the calculator and start the DataRad program. Press (CLEAR) to reset the program.

3. Prepare the DataRad program for this experiment.
   a. Select SETUP from the main screen.
   b. Select SET INTERVAL from the SETUP MENU.
   c. Select SET INTERVAL from the INTERVAL SETTINGS screen.
   d. Enter "50" as the count time interval in seconds. Always complete number entries with (ENTER).
   e. Select OK from the INTERVAL SETTINGS screen.
   f. Select EVENTS WITH ENTRY from the SETUP MENU.

4. Confirm that the source and monitor are positioned so they will not move, and so that there is enough space between them for ten layers of cardboard. Remove all but one layer of cardboard from between the source and monitor.

5. Select START to prepare for data collection. Confirm that all is in position, and then press (ENTER) to start the first counting interval. The interface will begin counting the number of beta particles that strike the detector during a 50-second count interval.

6. After 50 seconds have elapsed, DataRad will prompt you for a value. In the entry field that appears, enter "1", or the number of layers of cardboard.

7. Insert a second layer of cardboard between the source and detector. Be sure that the cardboard completely covers the source's "view" of the Geiger tube in the detector. Press ⌈ENTER⌋ to start the next count interval.

8. When the interval is complete, DataRad will again prompt for a value. Enter the new number of layers, or "2."

9. In the same way as before, add a layer of cardboard without moving the source or monitor, and press ⌈ENTER⌋ to start counting. When counting is complete, enter the number of layers of cardboard. Repeat this process until you have completed data collection for ten layers.

10. Press ⌈STO▸⌋ to end data collection and to display a graph.

DATA TABLE

Model equation	
Fitted equation with parameters	

ANALYSIS

1. Inspect your graph. Does the count rate appear to follow your model? How can you tell? After viewing your graph, press ⌈ENTER⌋ to return to the main screen.

2. Next you can fit an appropriate function to your data. To choose a function, select ANALYZE from the main screen. You will see a list of functions. Look for one that has the same mathematical form as your model. (Hint: which fit functions have an *x*, the horizontal axis variable, in the same special location as in your model equation?) Select an equation from the equation list. Record the parameters and the equation in your data table. Press ⌈ENTER⌋ to see a best-fit curve will be displayed on the graph. If your data follow the selected relationship, the curve should closely match the data.

3. Print or sketch your graph.

4. From the evidence presented in your graph, does the transmission of beta radiation through cardboard match that predicted by your model?

5. From the parameters of your fitted equation, determine the fraction *f* of beta rays transmitted, on average, by one layer of cardboard. Do not use your raw data to calculate the fraction, but instead use the information from your fitted equation. Hint: Remember that $A^{(Bx)} = (A^B)^x$.

EXTENSIONS

1. Use a longer counting interval so that you collect at least two thousand counts when no absorbing cardboard is in place. Is the agreement with the model any different? Try a much shorter count interval. How is the resulting graph different? Why?

2. Cosmic rays continue to strike the detector regardless of the absorbing cardboard. Measure the average background counts in one count interval, and correct your data for background radiation. Repeat the analysis.

3. Try other absorbers. For example, common household aluminum foil can be used in place of cardboard. You will need to experiment with the appropriate number of layers to use. You may want to add more than one (or five) layers at a time.

TEACHER INFORMATION

Radiation Shielding

1. Sources are available from a number of suppliers:

   - Spectrum Techniques, 106 Union Valley Road, Oak Ridge, TN 37830, (865) 482-9937, Fax: (865) 483-0473, www.spectrumtechniques.com.
   - Flinn Scientific, P.O. Box 219, Batavia, IL 60510, (800) 452-1261, Fax: (630) 879-6962, www.flinnsci.com.
   - Canberra Industries, 800 Research Parkway, Meriden, CT 06450, (800) 243-3955 Fax: (203) 235-1347, www.canberra.com.

2. Because the radiation monitors detect individual particle arrivals, Poisson statistics apply. The more counts that arrive in a counting interval, the better the precision. The standard error of a count of n is $n^{1/2}$, so do not be surprised to see considerable run-to-run variation in the many-layer points where n is only ten or twenty. Longer count intervals are required to achieve better precision.

3. This activity asks students to generalize the transmission through zero, one, two, three… absorbers of f^0, f^1, f^2, f^3, to the transmission through x absorbers: f^x.

4. It is critical that the geometry of the experiment remain constant as the absorbers are added. If either the monitor or the source is moved during data collection, the resulting run will probably be poor.

5. The final analysis question requires manipulating the fitted equation. Students who are weak in mathematics may need assistance with this step.

6. The analysis asks that the student choose an appropriate fit equation based on the mathematical form of the model. The model is an exponential function: $N = f^x N_0$. There are two exponential functions offered in Logger *Pro*. One is a base-10 exponential function of y = A*10^(B*x), or $Y = A10^{Bx}$. Another is a natural exponential function of y = A*exp(–C*x) + B, or $Y = Ae^{-Cx} + B$. The different base of the exponential function does not affect the shape of the function, but the natural exponential has the extra additive term of "$+ B$." Because the count rate is usually significantly higher than background, the additive term will have little effect on the fit. As a result, either function could be chosen for this experiment. The additive term does affect the fit slightly, however, so the exponential parameter will not be directly comparable in the two fits (aside from the base change). Since the base 10 exponential more closely matches the form of the model developed by students, it is the more natural choice, but either form can be used. The calculator does not offer the base 10 choice, and there is no additive term in the natural exponential fit.

7. The cardboard used for the sample data was cut from a standard cardboard shipping box. The transmittance will vary with type and thickness of cardboard.

8. The strontium-90 source used in this activity is a pure beta source. No gamma rays are emitted, so there is no confounding effect of differing absorption of gamma and beta radiation by the shielding material.

SAMPLE RESULTS

The model fits the experimental data well. The additional three points at lower left are measures of background radiation from cosmic rays; the background count rate was small compared to the count rate using the Sr-90 source.

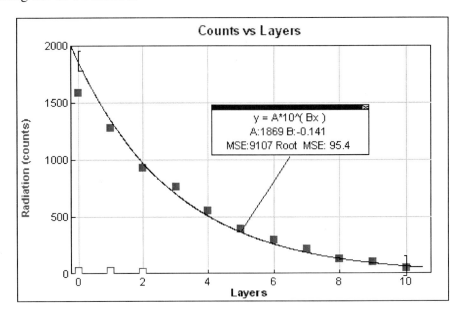

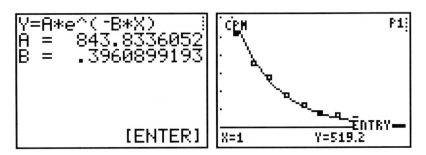

ANSWERS TO PRELIMINARY QUESTIONS

1. For the Student Radiation Monitor (black case with no meter), the clear window on the underside is the most sensitive region. For the Radiation Monitor (brown case with analog meter) the screen on the end is the most sensitive area.

2. Yes, the cardboard appears to shield the radiation monitor from the beta radiation of the Sr-90 source. Adding more layers of cardboard further reduces the count rate.

3. Graph is a decreasing function with additional layers of cardboard.

4. $N = N_0 f^x$, where f is the fraction of beta particles transmitted by one layer, and x is the number of layers.

5. Model function is also a decreasing function with additional layers x, since f is less than one.

DATA TABLE

Model equation	$N = N_0 f^x$
Fitted equation with parameters	$Y = 1870 \times 10^{(-0.141\,x)}$

ANSWERS TO ANALYSIS QUESTIONS

1. The count rate falls off rapidly with added absorbers. This is consistent with the model, which predicts reduced rates with increased numbers of absorbing layers.

2. See sample data.

4. Yes, the experimental data match the model fairly well, especially for the larger numbers of layers of absorbers. It appears that the simple multiplicative model does predict the transmission of radiation through matter.

5. (computer data) Using the base-10 fit, and noting that 10^{Bx} corresponds to f^x, we have $10^B = f$. So, $10^{-0.141} = 0.72$. One layer of cardboard (of the type used for the sample data) transmits 72% of the beta particles striking it.

   (calculator data) Noting that e^{-Bx} corresponds to f^x, we have $e^{-B} = f$. So, $e^{-0.39} = 0.68$. One layer of cardboard (of the type used for the sample data) transmits 68% of the beta particles striking it.

ANSWERS TO EXTENSIONS

1. For longer collection times, the total number of counts in each interval will be longer. As a result, the precision of each measurement will be greater. We would expect less scatter about the model line. For shorter collection times, the precision will be reduced and we would see more scatter about the model's function.

2. The sample data show three points below the main curve. These are background counts made with no source. The average was approximately 25 counts in each 50-second count interval. To correct for background radiation, subtract 25 from each of the data points collected using the source and the absorbers, and repeat the graphing and fits.

3. One layer of foil absorbs a much smaller fraction of the betas, so larger stacks of absorbers will be required. You may want to use ten layers of foil in place of each single layer of cardboard.

TI Graphing Calculators
and the DataRad Calculator Program

THE DATARAD AND VST APPS PROGRAM

The DataRad data collection program for the TI graphing calculator is a special purpose program for experiments using the Radiation Monitor and Student Radiation Monitor. DataRad works only with a Radiation Monitor connected to the DIG/SONIC1 port of the Vernier LabPro interface or the DIG/SONIC port of the Texas Instruments CBL 2 interface. DataRad is a single program, not a group. DataRad is not compatible with the original Texas Instruments CBL System. Compatible calculators include the TI-73, TI-83, TI-83 Plus, TI-84 Plus, TI-86, TI-89, TI-92, Voyage 200, and the TI-92 Plus.

The DataMate program included with the CBL 2 and LabPro interfaces does not support the Radiation Monitor or Student Radiation Monitor.

Users of the TI-83 Plus, TI-83 Plus Silver Edition, TI-84 Plus, and TI-84 Plus Silver Edition calculators can also use VST Apps. DataRad is one of the several accessory programs contained in the VST Apps application. As an application, VST Apps does not occupy data memory and so can be kept on the calculator alongside DataMate with no loss of data memory. We recommend that TI-83 Plus and TI-84 Plus users load VST Apps instead of the DataRad program.

Transferring the DataRad or VST Apps Program to a Calculator

DataRad and VST Apps are supplied on the CD included with this book and are transferred to the calculator using a TI-GRAPH LINK cable and TI Connect software. You can also download a copy of DataRad or VST Apps from www.vernier.com/calc/software/.

The DataRad program is available for each type of LabPro-compatible calculator. The versions are stored as individual programs called datarad.xxp, where xxp is one of 73p, 83p, 8xp, 86p, 89p, 9xp, and 92p for the various types of calculators.

The process for transferring these programs to your calculator is described here. You will need a computer, TI Connect software and a TI-GRAPH LINK cable.

Memory Note: The TI-73, and TI-83 calculators do not have enough memory to hold the DataMate program and its subprograms along with the DataRad programs. As a result, you must either delete DataMate and its subprograms or reset the calculator before loading these programs. You may need to delete some applications from the TI-83 Plus and TI-84 Plus calculators to have enough room for both DataMate and VST Apps.

Macintosh OS 9 Computers

    a. Connect the TI-GRAPH LINK cable to the USB or serial port of the Macintosh computer and to the port on the calculator.
    b. Turn on the calculator.
    c. Launch TI Connect.
    d. If no connection is automatically established, choose the type of calculator you have from the Connection menu. Choose the port used for the TI-GRAPH LINK cable (USB, Modem, Printer). Click Connect.

e. If a dialog offering to search for files on the Internet appears, click Cancel.

f. Locate the DataRad file or the VST Apps file. Drag the file to the TI Connect window and release. The program will be loaded on your calculator.

To confirm that the program is now in your calculator, press ⌈PRGM⌉ on the TI-73/83/83 Plus, 84 Plus, 86, or ⌈2nd⌉ [VAR-LINK] on the TI-89, TI-92, TI-92 Plus, or Voyage 200, and you should see the DATARAD program listed. If you loaded VST Apps on a TI-83/84, press ⌈APPS⌉ instead.

Macintosh OS X Computers

a. Connect the TI-GRAPH LINK cable to the USB port of the Macintosh computer and to the port on the calculator.

b. Turn on the calculator.

c. Locate the application TI Device Explorer, usually found in the folder TI Connect, inside the Applications folder. Launch TI Device Explorer. The list of items on the calculator is read by TI Device Explorer.

d. Locate the DataRad file or the VST Apps file. Drag the file to the TI Device Explorer window and release. The program will be loaded on your calculator.

To confirm that the program is now in your calculator, press ⌈PRGM⌉ on the TI-73/83/83 Plus, 84 Plus, 86, or ⌈2nd⌉ [VAR-LINK] on the TI-89, TI-92, TI-92 Plus, or Voyage 200, and you should see the DATARAD program listed. If you loaded VST Apps on a TI-83/84, press ⌈APPS⌉ instead.

Windows Computers

Using the Windows version of TI Connect, follow these steps to send the program DataRad or the application VST Apps to the TI calculator.

a. Connect the TI-GRAPH LINK cable to a serial or USB port of your computer and to the port on the calculator.

b. Turn on the calculator.

c. Launch TI Connect on your computer. Click on Device Explorer.

d. Locate the DataRad file or the VST Apps file. Drag the file to the TI Device Explorer window and release. The program will be loaded on your calculator.

To confirm that the program is now in your calculator, press ⌈PRGM⌉ on the TI-73/83/83 Plus, 84 Plus, 86, or ⌈2nd⌉ [VAR-LINK] on the TI-89, TI-92 or TI-92 Plus, and you should see the DATARAD program listed. If you loaded VST Apps, press ⌈APPS⌉ instead.

DATARAD PROGRAM DETAILS

Start DataRad

Use the black link cable to connect LabPro or CBL 2 to the TI graphing calculator. Firmly press in the cable ends. Turn on the calculator. Follow these steps to start the DataRad program on your calculator:

TI-73 and TI-83 Calculators

Press ⌈PRGM⌉ and then press the calculator key for the *number* that precedes DATARAD (usually ⌈1⌉). Press ⌈ENTER⌉ and wait for the main screen to load.

TI-86 Calculators

Press [PRGM] and then press [F1] to select <NAMES>, and press the menu key that represents DataRad. (<DATAR> is usually [F1]). Press [ENTER], and wait for the main screen to load.

TI-89, TI-92, or TI-92 Plus, Voyage 200 Calculators

Press [2nd] [VAR-LINK]. Use the cursor pad to scroll down to "datarad", then press [ENTER]. Press [ )] to complete the open parenthesis that follows "datarad" on the entry line and press [ENTER]. Wait for the main screen to load.

TI-83 Plus and TI-84 Plus Calculators

Press [APPS] and then press the calculator key for the *number* that precedes VST APPS. Once the main screen loads press the *number* that precedes DATARAD.

Main Screen

The main screen of DataRad is shown at the right. The top half of the screen shows the Radiation Monitor reading, the current data collection mode, and the status of background correction. The portion below the double bar displays the menu options.

```
DIG1:RADIATION(CPM)
MODE:TIME GRAPH- 60

BACKGROUND CORRECT:NONE
PRESS [+] TO TURN ON
━━━━━━━━━━━━━━━━━━━
1:SETUP      4:ANALYZE
2:START      5:QUIT
3:GRAPH
```

Press [+] to toggle the background correction mode on and off.

Below the double bar you will find a varying number of menu items, depending on the mode.

SETUP

Select SETUP to select the data collection mode, counting interval length, and background correction status. Additional information on the various setup options follows in the Setup Menu section.

START

Select START to begin data collection.

GRAPH

Select GRAPH to see the current graph. If no data have been collected, selecting GRAPH will end the program.

ANALYZE

Select ANALYZE to see a list of curve fitting, modeling, and range selection functions.

```
     ANALYZE OPTIONS

1:LINEAR CURVE FIT
2:EXPONENT CURVE FIT
3:POWER CURVE FIT
4:ADD MODEL
5:SELECT REGION
6:RETURN TO MAIN SCREEN
```

QUIT

Select QUIT to leave the DataRad program.

Setup Menu

SET INTERVAL

Select SET INTERVAL to see the current interval length in seconds. The DataRad program counts for this interval before reporting a result. From the Interval Settings screen, select SET INTERVAL to change the current setting, or select OK to return to the Setup Menu.

```
         SETUP MENU
1:SET INTERVAL
2:BACKGROUND CORRECTION
3:TIME GRAPH
4:RATE/HISTOGRAM
5:EVENTS WITH ENTRY
6:SINGLE INTERVAL
7:RETURN TO MAIN SCREEN
```

BACKGROUND CORRECTION

Background correction allows you to display count rates corrected for the presence of any background radiation. Background correction must be set before an experiment is performed, and the interval to be used in the experiment must be set before setting up background correction.

```
   BACKGROUND CORRECTION
1:PERFORM NOW
2:RESET TO NONE
3:RETURN TO SETUP SCREEN
```

Perform Now

Select PERFORM NOW to measure the background count rate and set up DataRad to correct for this count rate. You will be prompted for the number of intervals over which you want DataRad to count background events. Enter an integer (usually five or so), remove all radiation sources from the vicinity of the Radiation Monitor, and press [ENTER] to begin counting. After the entered number of intervals has passed, DataRad will display the average count rate in cpm (counts per minute) by which all following measurements will be reduced. In order to prevent later curve fits from failing in time and event-based data collection modes, all zero or negative corrected count rates will be changed to a count rate of 0.1 cpm. A zero or negative count rate can occur if the measured count rate is very low, and statistically happens to be less than or equal to the previously determined background correction.

Reset to None

Select RESET TO NONE to clear a previously determined background correction. You can also temporarily disable background correction from the main screen by pressing [+].

Return to Setup Screen

Select RETURN TO SETUP SCREEN to return to the Setup screen.

TIME GRAPH

Select TIME GRAPH to configure DataRad to collect count rate data as a function of time. For example, a lifetime measurement experiment would use this mode. A screen showing the current interval in seconds, the current number of samples to be collected, and the product of the two (the experiment length) will be shown. Select OK to accept the settings, and select CHANGE TIME SETTINGS to adjust either the interval or the number of samples. Return to the main screen and select START to begin data collection.

```
   TIME GRAPH SETTINGS
INTERVAL:          5
NUMBER OF SAMPLES: 12
EXPERIMENT LENGTH: 60

1:OK
2:CHANGE TIME SETTINGS
```

RATE/HISTOGRAM

Select RATE/HISTOGRAM to set the data collection mode to Rate/Histogram. You will then return to the main screen. Begin data collection by selecting START. Data collection will begin, and the screen will display the number of intervals collected, the most recent count rate, and the average count rate over all intervals. Press [STO] to end data collection. A histogram will be displayed. You may scan across the histogram using the cursor keys.

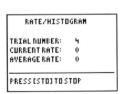

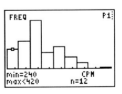

Once you are done with the histogram, press [ENTER]. You will be prompted to adjust the number of bins for the histogram if desired. Select YES or NO. If you select YES, at the prompt enter a number between one and 45. The rescaled histogram will be displayed.

After selecting NO, the average, standard deviation, minimum, maximum and number of intervals (trials) will be shown.

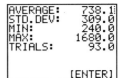

EVENTS WITH ENTRY

Select EVENTS WITH ENTRY when the experiment requires count rates as a function of some parameter other than time. This parameter could be distance, number of layers of shielding, or another parameter under your control. In this mode DataRad will count for one interval and then prompt you for a numeric entry. After making that entry, data collection is paused so that you can reconfigure the apparatus for the next data collection condition. You then press [ENTER] to recommence counting, after which DataRad again prompts for a numeric entry. Data collection can be ended after any numeric entry by pressing [STO]. After data collection has ended, DataRad displays a graph of count rate *vs.* the entered values.

SINGLE INTERVAL

DataRad can be set to count for a single interval. After setting the desired data collection interval, select the single interval mode. From the main screen select START. The screen shows count rates in counts per interval and counts per minute, both raw and corrected by any previously configured background correction.

Analyze Options

ANALYZE OPTIONS allows the user to perform basic data analysis within the DataRad program.

LINEAR CURVE FIT

A straight-line least squares fit of the form $Y=A*X+B$ is performed on previously collected data. You may want to first select a portion of your data using the SELECT REGION option described below. After the fit parameters are displayed, a graph of the data with the fitted curve is shown. Press [ENTER] to continue.

EXPONENT CURVE FIT

An exponential fit of the form $Y=A*EXP(-B*X)$ is performed on previously collected data. You may want to first select a portion of your data using the SELECT REGION option described below. After the fit parameters are displayed, a graph of the data with the fitted curve is shown. Press [ENTER] to continue.

POWER CURVE FIT

A power law fit of the form $Y=A*X^B$ is performed on previously collected data. You may want to first select a portion of your data using the SELECT REGION option described below. After the fit parameters are displayed, a graph of the data with the fitted curve is shown. Press [ENTER] to continue.

ADD MODEL

To use this feature you must enter an equation in the Y1 equation variable prior to running the program. The equation can contain up to five adjustable parameters A, B, C, D and E. You may collect data, leave the program to enter the equation, and then restart DataRad without losing the data.

The ADD MODEL feature allows the user to superimpose a manually controlled function on the data. A user-entered equation with up to five parameters is graphed. For example, if the data are to be compared to an inverse-square function, the Y1 equation variable could be set to A/X^2. The parameter A is then adjusted after selecting ADD MODEL. After selecting the parameter A and entering a value, the data and the equation using that parameter are graphed. With certain values of the parameters the equation will not be visible on the data graph. Press ENTER to return to the MODEL MENU. Usually you will want to adjust the parameter several times to optimize the fit of the equation. Allowed parameters are A, B, C, D and E.

SELECT REGION

The SELECT REGION option allows you to delete data to the left and right of the desired region. You may want to use this function before performing a curve fit. To use, select SELECT REGION. You will see your data graphed with a left-bound cursor. Move the cursor to the desired left bound, and press ENTER. Then move the right cursor to the desired right bound, and press ENTER. Data outside the indicated region will be permanently deleted.

RETURN TO MAIN SCREEN

Select this to return to the main screen of DataRad.

Transferring Data to a Computer

If you wish to analyze the data you collected with a TI graphing calculator on a computer or if you want to print the data, you will need to use the TI-GRAPH LINK cable to transfer the data to Logger *Pro* or Graphical Analysis computer software. The following procedures are for use with TI-73, TI-73 Explorer, TI-83, TI-83 Plus, TI-83 Plus Silver Edition, TI-84 Plus, TI-86, TI-89, TI-92, TI-92 Plus, and the Voyage 200 calculators.

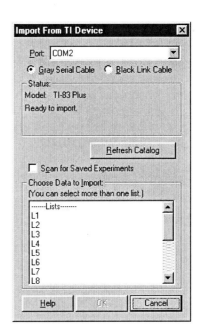

1. Connect the TI-GRAPH LINK cable to a free serial or USB port of the computer and to the TI calculator.

2. Turn on the calculator.

3. With Logger *Pro* or Graphical Analysis running, pull down the File menu and choose Import From → TI Calculator or LabPro/CBL... or click the 📲 button from the toolbar. The dialog box to the right should appear.

4. From the pull-down menu, choose USB port or serial port (COM 1-4 on a PC, modem or printer on a Macintosh) to which the TI-GRAPH LINK cable is connected.

5. If you are using a PC serial cable, identify whether it is a gray or black cable.

6. Click on the Scan for Calculator button. The calculator model you are using should now be identified, and you should see a message, "Ready to Import."

7. Click on the data list(s) that you want to import. (To select more than one list on a Macintosh, hold down the ⌘ key while you click).

8. Click OK to send the data lists to the computer. The data will appear in columns in the data table. They will be labeled with the simple list names from the calculator. If you want to re-name them or add units, double-click on the column heading in the data table and enter new labels and units.

Using the CD

The CD located inside the back cover of this book contains one folder of calculator programs and four folders containing word processing files.

▢ **Calculator Programs** — Required for all experiments in this book if they are to be done using a Vernier LabPro or TI CBL 2 and a Texas Instruments graphing calculator.

▢ **Nuclear Rad Calculators Word** — These are word-processing files for each of the six student calculator-based experiments in this book. DataRad calculator software (found in the Calculator Programs folder) is required to perform the experiments.

▢ **Nuclear Rad Computers Word** — These are word-processing files for each of the six student computer-based experiments in this book. Logger *Pro* version 3.3 or newer is required to perform the experiments.

▢ **Nuclear Rad Handhelds Word** — These are word-processing files for each of the six student Palm OS handheld-based experiments in this book. DataPro version 1.5 or newer is required to perform the experiments.

▢ **Word Files for Older Book** — These are the word-processing files for each of the six student computer-based experiments in the book *Nuclear Radiation for Computers and Calculators, First Edition.* They use Logger *Pro* 2 computer software with LabPro or ULI.

Calculator Programs

Refer to Appendix A for information on using the DataRad calculator program files.

Nuclear Radiation Word-Processing Files

The Word files provide a way for you to edit student experiments to match your lab situation, your equipment, or just your style of teaching. They contain all figures, text, and tables in the same format as printed in this book.

To use these files, start Microsoft Word or another word processor capable of importing Microsoft Word files. Open the file of your choice from the calculator or computer folder. Files can be opened directly from the CD or first copied onto your hard drive. Files are in Word 2002 format, which can be opened by many word processors.

Experiment   File Name

Computer

1 α, β, and γ .... 1 Alpha

2 Distance and Radiation .... 2 Distance

3 Lifetime Measurement .... 3 Lifetime

4 Counting Statistics .... 4 Statistics

5 Background Radiation Sources .... 5 Background

Vernier Products for Nuclear Radiation

Software

Hardware

All software and laboratory interfacing hardware required for the experiments contained in this book is available from Vernier Software & Technology. The purchase of Vernier programs includes a site license that permits you to make as many copies as you wish for use in your own school or college department. You may also use Logger *Pro* on networks within your school at no extra cost. You may want to purchase a set of the hardware for each computer used in your lab. Vernier Radiation Monitors and Student Radiation Monitors can be used with the LabPro on Macintosh or PC computers. The Radiation Monitors can also be used with the LabPro with a Texas Instruments graphing calculator or Palm OS handhelds.

Vernier LabPro

Vernier LabPro provides a portable and versatile data collection device for any class studying science. A wide variety of Vernier probes and sensors can be connected to each of the four analog channels and two sonic/digital channels. LabPro is connected to a computer using a serial or USB port, to a TI graphing calculator, or to a Palm OS handheld. Data collection with LabPro and a computer is controlled by Logger *Pro* software.

Included with the purchase of a LabPro (order code LABPRO) is a Voltage Probe, computer cables (USB and serial for Macintosh and Windows), calculator cradle, short calculator link cable, user's manual, and AC power supply.

CBL 2

Texas Instruments CBL 2 is also a portable and versatile data collection device. A wide variety of Vernier probes and sensors can be connected to each of the three analog channels and one digital/sonic channel of the CBL 2. The CBL 2 is connected to a TI calculator through the port found on the bottom edge of the calculator. Data collection with the CBL 2 is controlled by the DataMate data-collection program that comes stored on the CBL 2. Because the CBL 2 is battery powered, it can be taken out of the classroom to monitor data in the field. Within the classroom, the CBL 2 provides a low-cost alternative to the use of computers.

Included with the purchase of a CBL 2 is a calculator cradle, short calculator link cable, temperature probe, light probe, voltage probe, experiment workbook, user's guidebook, batteries, and carrying case (order code CBL2).

Which Radiation Monitor?

Either the Student Radiation Monitor or the Radiation Monitor may be used for the six experiments in this book. The Student Radiation Monitor is less expensive, and is slightly more sensitive than the Radiation Monitor. The Student Radiation Monitor cannot detect alpha particles, however, so the first experiment cannot be performed for alpha sources. The Radiation Monitor also can be used without a computer with its built-in ratemeter.

Item	Price and Order Code
Student Radiation Monitor	SRM-BTD
Radiation Monitor	RM-BTD

Software for Nuclear Radiation

Logger *Pro*

Logger*Pro* software is the data-collection software for the *computer* experiments in this lab manual (when using a LabPro to collect data). It can also be used to import data collected on a calculator or a Palm OS handheld. This award-winning program is just what your students need to plot graphs of their experimental data. The graphs produced follow standard graphing conventions and can include point protectors, a background grid, error bars, or linear regression best-fit lines. Graphical Analysis can be used to create modified versions of a graph. You can modify the data on either axis by raising it to any power, taking the log, or using a trig function. This can be very useful in finding the relationship between variables. Automatic and manual curve fitting, calculated columns, histogram graphs, and text windows are also available. Since all of our software programs include a site license, you only need to order one copy of Logger *Pro* for your school or college department. Logger *Pro* is for both Mac and Windows (order code LP).

Graphical Analysis

This program contains all of the graphing features of Logger *Pro* but without data collection capabilities. It can be used to import data from a TI graphing calculator or a Palm OS handheld for further analysis and printing.